the seed underground

also by janisse ray

Ecology of a Cracker Childhood

Wild Card Quilt: Taking a Chance on Home

Pinhook: Finding Wholeness in a Fragmented Land

A House of Branches: Poems

Drifting into Darien:
A Personal and Natural History of the Altamaha River

praise for *the seed underground*

"Traveling about the country to introduce us to some of her devoted fellow seed savers, Janisse Ray teaches us more than we thought we needed to know about seeds: how remarkable they are, why they need saving, how to save them, and how many of them—each holding the future of some particular plant—have been lost and are being lost to our indifference. But in a world where everything we love—including seeds—seems to be under threat, Ray ultimately offers us hope. 'Everything the seed has needed to know is encoded within it,' she assures us, 'and as the world changes, so it will discover everything it yet needs to know.' A poetic, and always hopeful, book."

—Joan Gussow, author of *Growing, Older* and *This Organic Life*

"What a dream of a book—my favorite poet writing about my favorite topic (seeds) and the remarkable underground network of growers who are keeping diversity alive on the face of this earth while putting delicious food on our tables! If books can move you to love, this one does."

—Gary Paul Nabhan, coauthor of *Chasing Chiles* and editor of *Renewing America's Food Traditions*

"If I get to feeling a little blue about our prospects, I'm liable to reach down one of Janisse Ray's books just so I can hear her calm, wise, strong voice. This one's my new favorite; a world with her in it is going to do the right thing, I think."

—Bill McKibben, founder of 350.org

"This is an important book that should be required reading for everyone who eats. Big biotech companies are patenting and privatizing seeds, making it illegal for farmers to retain their own crops for replanting. In a series of engaging and lyrical profiles, Ray shows that by the simple and pleasurable act of saving seeds we can wrest our food system from corporate control."

—Barry Estabrook, author of *Tomatoland: How Modern Industrial Agriculture Destroyed Our Most Alluring Fruit*

"This is an unmatched treasure trove of information. . . . *The Seed Underground* is an excellent choice for readers seeking a depiction of the current critical situation in farming all in one, easy-to-read book."

—Gene Logsdon, author of *A Sanctuary of Trees* and *Holy Shit*

the seed underground

a growing revolution to save food

JANISSE RAY

CHELSEA GREEN PUBLISHING

WHITE RIVER JUNCTION, VERMONT

Project Manager: Patricia Stone
Developmental Editor: Brianne Goodspeed
Copy Editor: Eric Raetz
Proofreader: Susan Barnett
Indexer: Lee Lawton
Designer: Melissa Jacobson

Printed in the United States of America
First printing June, 2012
10 9 8 7 6 5 4 3 2   12 13 14 15 16

Our Commitment to Green Publishing
Chelsea Green sees publishing as a tool for cultural change and ecological stewardship.
We strive to align our book manufacturing practices with our editorial mission and to
reduce the impact of our business enterprise in the environment. We print our books and
catalogs on chlorine-free recycled paper, using vegetable-based inks whenever possible.
This book may cost slightly more because it was printed on paper that contains recycled
fiber, and we hope you'll agree that it's worth it. Chelsea Green is a member of the Green
Press Initiative (www.greenpressinitiative.org), a nonprofit coalition of publishers,
manufacturers, and authors working to protect the world's endangered forests and
conserve natural resources. *The Seed Underground* was printed on FSC®-certified paper
supplied by Thomson-Shore that contains at least 30% postconsumer recycled fiber.

Library of Congress Cataloging-in-Publication Data
Ray, Janisse, 1962–
  The seed underground : a growing revolution to save food / Janisse Ray.
    p. cm.
  ISBN 978-1-60358-306-0 (pbk.) — ISBN 978-1-60358-307-7 (ebook)
1. Seeds—United States. 2. Vegetables—Heirloom varieties—United States.
3. Fruit—Heirloom varieties—United States. 4. Vegetables—Seeds—United States.
5. Fruit—Seeds—United States. 6. Gardening—United States—Anecdotes. I. Title.

SB117.3.R39 2012
631.5'21—dc23
                                    2012014556

Chelsea Green Publishing
85 North Main Street, Suite 120
White River Junction, VT 05001
(802) 295-6300
www.chelseagreen.com

For Wendell Berry
No monument would be tall enough.

When they want you to buy something,
they will call you. When they want you to die for profit,
they will let you know. So, friends, every day
do something that won't compute.

—Wendell Berry, "Manifesto:
The Mad Farmer Liberation Front,"
from *The Country of Marriage*

contents

CONTENTS

preface

I AM STANDING under the saddest oak that ever was. A young man who as a child climbed this very tree has died, fallen from a balcony during a party. For his memorial service there are no pews, altar, or casket. A circle of friends and family congregate in the yard of his grieving mother, who is my friend.

One tattooed man of about twenty-five remembers how his buddy helped him through bouts of depression. Another young man behind me steps forward to speak, stumbles, and throws his arm around my waist. His kind eyes are bloodshot and he breathes out small alcoholic clouds. He steadies himself and delivers a short poem, words scribbled large in blue ink on a sheet of paper, like hieroglyphics. No doubt the gibberish sounded wise when it was written.

This man is the age of my son, as could be any of them, young people searching for beauty and meaning, struggling to understand the events of the epoch. They are so young to be so familiar with grief. As we stand in the Florida Panhandle sunshine, a broken pipe from an exploded rig leaks millions of gallons of oil into the Gulf of Mexico. Already sharks and fish are washing up dead along the coast one hundred miles away. We could be holding a funeral for the Gulf, or for the climate, or for any number of things.

I feel around me a cavernous hopelessness. But I do not feel hopeless.

Many systems that we collectively have been living amid and on which we rely appear to be failing. The easiest thing to do is to give up. But so much needs to be done; every mind and body is crucial for putting new systems in place. We need positive contributions. We don't need people to drop out.

On the way home from the service I was listening to bluegrass when I heard this question: *What will you be building when you are called away?*

This book is for everyone, but it is especially for young people, in hopes that, given all the bad, you start building. Not skyscrapers or oil rigs, but lives that make sense, that contribute to a lighter, more intelligent, more beautiful way of living on the earth, lives that are lived as far outside and beyond corporate control as possible.

That in doing so you find meaning.

That you attain higher planes before drugs or alcohol enslave you.

That you inhabit the country of love.

That you find happiness.

Here in the country, on a little farm in southern Georgia, I am building a quiet life of resistance. I am a radical peasant, and every day I take out my little hammer, and I keep building.

Seeds are only a small part of life. But they represent everything else. All our relations.

introduction

"I have great faith in a seed."
—HENRY DAVID THOREAU

As THE PRICE of gold and silver rises and the value of paper money nose-dives, the most priceless commodity we humans own goes largely unnoticed. When presented with this capital, most people have no place for it, do not know what to do with it, and do not value it. Thus it winds up in the trash.

Yet far beyond Wall Street, beyond the supersized stores of a disappearing industrial world now past its heyday, these precious goods travel from hand to hand in one of the most interesting economies in the world.

One morning a bit of this economy passes to me. The sun knocks around in the eastern sky, spilling yellow light everywhere, when I leave our farmhouse. I run along the fence and four cows gallop with me, acting more like dogs than cows. They are named for activists— Emma, Che, Geronimo, and Amy. When the pasture ends, they stop and I keep running.

Across the dirt road, rye in a field has gone to seed, and the sunlight catches and hangs there in its awns until the rye looks like an array of tiny hedgehogs. The mailbox is a mile away and I jog through a corridor of immense trees, their leaves new-minted and bright as limes. Toadflax blooms in pockets of bare sand. A trumpet vine encircles one pine, transforming it into a column of red blossoms.

I gather yesterday's mail from our box on Old River Road and am running back when I meet my neighbor, puttering along in his faded black truck. Not that many years ago he would have been on a horse, and suddenly I miss something I have never known. My neighbor rolls to a stop.

"Howdy to you," he says. Mr. Stanley is a retired nurseryman, about seventy. Just after my family moved in, he came by to let us know that a certain camellia growing by our porch steps was rare, begging us not to cut it down.

"Good morning!" I say to him now.

"Mrs. Stanley and I were talking about you last night," he says. In his shirt pocket is an open pack of long, dark cigarettes.

"All good, I hope."

"Oh, yes." He looks off up the road. "About 1880 or 1890," he says, "my great-grandfather, Joe Stanley, crossed three kinds of corn and developed his own variety. Of course, that was back in the days before hybrids and so forth. We've kept it growing in my family ever since." He looks back at me. "Last night Mrs. Stanley and I were thinking that we could share our seeds with you."

I want to ask him to turn off the truck, which is idling, vaporizing carbon dioxide into the atmosphere.

"You come from a long line of plantsmen, don't you?" I ask. "Is it sweet corn?"

"No," he says, "for grits and meal." Mr. Stanley gestures with his lighter. "In fact, that cornmeal we gave you at Christmas was from this. It's a white corn."

"That was delicious."

"Well, we'd be glad to share the seeds with you."

"I'd love to try them, Mr. Stanley. I was saying this morning that it's time to plant corn."

"Call me Howard," he says.

"I'll try."

"I better get on," he says. "We're repairing the split-rail fence today."

"By the way, Howard, does the corn have a name?"

He makes a ceremony of clasping the steering wheel with both hands and looking at me. The morning sun could be maple syrup pouring through pine trees. "Yes, ma'am," he says, humble as can be. "It's Stanley corn."

If you haven't heard what's happening with seeds, let me tell you. They're disappearing, about like every damn thing else. You know the story already, you know it better than I do, the forests and the songbirds, the Appalachian Mountains, the fish in the ocean. But I'm not going to talk about anything that's going to make us feel hopeless, or despairing, because there's no despair in a seed. There's only life, waiting for the right conditions—sun and water, warmth and soil—to be set free. Every day millions upon millions of seeds lift their two green wings.

I've waited for those same conditions, and now I've found them. I believe that you will too. We are in the springtime, in the garden, winging into a new era, the Ecozoic era, and you and I have just alighted. Welcome.

All my life I dreamed of being a farmer. My mother had been glad to leave the farm, where I spent many Saturdays with my grandparents Arthur and Beulah, whose children one by one had moved away to the big Southern cities of Jacksonville, Orlando, Chattanooga. Deep in my psyche are my grandfather's mules, my grandmother's chickens, fields of vegetables and sprawling watermelon vines, full corncribs. During my preschool years, my grandmother milked a cow. Then there was the Farmall-A tractor and bird guano fertilizer, and after my grandfather died, when I was six, subsidized tobacco and Roundup weedkiller, monster combines and terrible erosion and the invasion of privet. The cane grinder was sold, the smokehouses fell, the last hen wasn't even eaten. Grandmama sent the milch cow to the livestock auction. I remember her final pea-patch.

On that same farm, the one I roamed as a child eating crabapples and muscadines, pomegranates and sand pears, now the story is Roundup-resistant pigweed growing among rows of genetically modified (GM) soybeans in fields leased to chemical cultivators. The fencerows are bulldozed, demolishing the plantings of wax myrtle and wild cherry accomplished by mockingbirds and cardinals. Fences are yanked out and the farmers are crowding right up to the road, since the field has to get bigger at all costs. The sassafras tree my grandfather so carefully skirted with his harrows is dead and gone.

Any one of us middle-aged Americans could be the poster child for the story of agriculture in the United States, one that began with working farms; farm animals; seed saving; land-based, subsistence economies; farming children. And, poof, all that was gone, brushed aside casually.

It happened so quickly. It left me doodling pitchforks in college astronomy and world civ notebooks.

I'm back. Not on my grandmother's farm, but somebody's grandmother's farm. It has forty-six acres in pasture, field, and woods. It has a house built in 1850 by a man whose brother operated a sawmill on Slaughter Creek, which collects water off the fields of Reidsville, Georgia's prison farm, and delivers it to the Altamaha River.

Our gardens are gridworks of raggedly rectangular raised beds. The gardens are fenced to keep the barnyard fowl—including the crazy guineas, which scuttle around like boats on legs hollering their mad, prehistoric calls—from scratching up every seed we plant. We have a pig or two, a few goats and sheep, some chickens, some turkeys and ducks.

Many people still alive today have seen the entire process of American ag: the function, the falling apart, the rise of big chemical, and now the coming back. We are witnessing in agriculture a revolution, a full circle.

Except it's not a circle. We are not returning to where we were. With some of the old knowledge intact and armed with fresh knowledge, we are looping forward to a new place. And we're coming there different. We are coming better prepared. We're coming educated. Girls as well as boys are coming. We're coming as greenhorns, but we're coming together.

We're coming knowing that failure is not possible. To not fail, we desire to understand everything we can about the cycle of life. We plumb the depths of industrial empire. We can no longer believe in false magic, that whatever we hanker for will be available to us, as it has been for most of our lives, whenever we wanted it, that it will appear magically in stores and restaurants as long as we have money wadded in our pockets.

At no time in our history more than now have Americans been more knowledgeable and more concerned about what we eat. We have watched our food systems deconstructed in front of our eyes. In a way, the farmer in all of us has roused. We understand organic, that food grown without chemicals is healthier for us and the earth. We understand local, that food grown closer to home is healthier and helps solve the climate crisis as well. Now we come to the landscape of American agriculture with a fresh realization: We do not have control of seeds, which are the crux of our food supply. When we dig deeper, we realize that our seed supply is in crisis and therefore our food is in crisis. A tragedy of corporate robbery is being acted out on a world stage, except this is not a drama with us in the audience getting to go home afterward. This is real.

The time has come to understand food at its most elemental.

I want to tell you about the most hopeful thing in the world. It is a seed. In the era of dying, it is all life. Every piece of information necessary to that plant for its natural time on earth is encoded, even though the world is changing and new information will be needed. But we don't know what is in a seed; its knowledge is invisible, encased, secret. A seed can contain any number of surprises. A seed can contain a whole tree encrypted in its sealed vault. Even with climate change there will be seeds that have all the wisdom they and we need.

Seeds are everywhere yet nobody thinks of them, tiny bundles waiting to be opened. Most any seed is small enough to hide in a locket, and some are so small they get latched between two teeth. They are on our forks, between the cracks of sidewalks and cobblestones of streets, under trees, on trees, in the produce section of the supermarket. They are in the wind, rolling along the desert, in our hair, in the fur of animals, in our dogs' feet.

As a mother, I came to fall in love with young people, having opened my home and dining table and toy box to so many of my son's friends over the years. Having opened my heart. Now Silas, in college, thinks I'm radical enough not only to introduce to his friends, but to walk down a street holding my hand. It is an ecstatic feeling, beyond joy, that a young man would feel this way about a parent. My job, too, as a writer, ferries me to many university campuses, where I find myself engaged in honest and deeply transformative conversations with young thinkers who understand very clearly what is not working. Tattooed arms and studs do not scare me, nor do hip boots with miniskirts or low-rider pants. I am not afraid of nudity, nor long hair, nor unshaven armpits. All of this is part of the story of belonging. I accept you.

There was a time when I thought nothing was sexier than learning to plow with a team of oxen, then treading the furrows, dropping kernels. I have reached the age, fifty, where I see my own life force ebbing away and I want to empower others, especially young people, as I have been strengthened. Plus there's love. The story is about love. There is nothing else for it to be about. I love you. I love you more because you love even the dirt. Or especially the dirt.

Even though I may not know you, I have fallen in love with you, you who understand that a relationship to the land is powerful; who want that connection; who want authentic experiences; who want a life that has meaning, that makes sense, that is essential. And I am writing for you. You. This story is for you. This is not a textbook on seed saving. I am looking to inspire you with my own life.

more gardens, less gas

THE WOMAN WHO ANSWERED my knock didn't look like a revolutionary. She was slim, in blue jeans and hyacinth turtleneck. Sporty reading glasses hung from her neck.

"Right on time," she said.

I smiled. "For once."

When I decided to learn as much as I could about seeds, I was directed to a village in central Vermont where a woman lives—a quiet, under-the-radar revolutionary, I was told—who understands some things I'm trying to understand.

She invited me inside, into a sparkling and artful kitchen. The walls were red, the stove green, the counters blue. On a woodstove rested a pan filled with seed heads I did not recognize. The woman followed my eyes. "Leeks," she said.

Meet Sylvia Davatz, radical American gardener. Somewhere in her well-kept home in the forested hills of central Vermont is a seed collection of plant varieties salvaged from the dustbins of history. "I'm the Imelda Marcos of seeds," she laughed. "I have a thousand varieties in my closet!"

She asked if I wanted to go right away to look at the gardens, since the temperature was still cool. I said that I did, and I followed Sylvia out a patio door and into a backyard full of riches, better than stocks and bonds, silver-plated with dew. As we wandered through her garden, she talked about the state of the world; afterward, as I mulled over her thoughts, one line kept turning like plow-dirt in my head.

"The system is so broken," she said. "Not only broken, but destructive and self-destructive." By "system," I figured she meant the agricultural or food system. Maybe she meant the entire political system. But I didn't ask, I just listened. "I see in activism a kind of futility," she said, brown eyes sincere. "The real power is in doing. The real power is in making the system irrelevant. That means nonparticipation in the existing broken system."

Sylvia didn't know I was an activist. I organized rallies to protest the climate crisis. I dressed in a penguin costume and waved an END CLIMATE CHANGE sign at a gallery walk. Wearing wetsuits, three friends and I tubed down Vermont's West River one January to highlight the fact that the water wasn't frozen a foot thick, as it should have been. We hung a WHERE'S WINTER? banner from the Dummerston covered bridge. I watched two friends get Tasered as we protested a proposed truck stop: MORE GARDENS, LESS GAS. I petitioned and wrote letters to editors and called politicians. A couple of times I got arrested.

Sylvia wasn't protesting anything in her peaceful garden: "What I am doing is making a broken system irrelevant."

Why would I call a quiet gardener and simple seed saver a revolutionary? First you must understand what is happening with seeds.

Let's start more than one hundred million years ago, around the time flowering plants appeared, in the great inflorescence of earth, as the late theologian Thomas Berry called it. For millions of years, flowering plants evolved, diversifying and developing sophisticated mechanisms for growth and reproduction. Humans first appeared on earth much later, a couple hundred thousand years ago. For our entire tenure on this planet, we have been surrounded by flowers, by the pollinators that evolved to tend the flowers, and by the subsequent fruits and seeds that those flowers produced. We are truly Flower Children.

A history of civilization is a history of seeds. Thousands of years ago we earthlings, hunter-gatherers, began to use flowers to our advantage. We started to experiment with growing food rather than simply chasing it or wandering around looking for it and so inaugurated plant domestication and evolution to the agrarian cultures that bore us all. Seeds, and the development of varieties, both allowed and encouraged our ancestors to settle down and become agricultural.

When humans roved from Asia across the Bering land bridge to what we know today as the continent of North America, they carried seeds with them. Likely one was the gourd, which originated in Africa, homeland of modern humans. It was transported to Asia in the original migrations and from East Asia was brought to North America.

Cultivation has at its root the simple word *cult*—from the Latin *cultus*, to care for. For thousands of years—at least 12,000, but perhaps many more—humans have engaged in plant domestication, developing and caring for progenitors of the crops familiar to us today. (The date of agriculture's genesis is highly contested. Some archaeologists believe that agriculture was happening 23,000 years ago, even in North America. However, in his book *Africa: A Biography of a Continent*, journalist John Reader estimates that the manipulation of food crops could have begun 70,000 years ago. He bases his dates on both carbonized deposits of native tubers and roots in cave sites in South Africa and on a definition of agriculture as a process of manipulating the distribution and growth of plants so that greater quantities of their edible parts are available for harvesting and consumption.)

Plant domestication conceivably happened like this: Out walking along some river, perchance the braided Nile, some long-ago human would discover a strange fruit and would taste it. She would not die. The tribe might save the seeds and begin to grow the plant in spots where they paused during a summer sojourn. Then someone would discover that a certain crop had a particularly large fruit, and that plant would be specially guarded and its seed saved to plant the following spring. A plant with a large fruit might be crossed with a plant with a sweet fruit. And on and on.

Corn, for example, was not discovered in some far valley, pendent with plump, sweet ears. No—corn was developed. The genetic ancestor of corn is teosinte, a large grass with multiple stalks whose ear is not even as long as a jackknife and is simply a line of triangular seeds linked to a stem. Its kernels are so hard they will break your teeth.

Modern corn is most genetically kin to a teosinte that still grows in the Balsas Valley of Mexico. Mesoamerican women planted teosinte. From it they chose the longest ears with the fattest kernels, and were jubilant when they crossed plants whose harvest eclipsed all prior. This was happening about 9,000 years ago in southwestern Mexico, long before European contact.

Over several centuries humans transformed teosinte into the plant we know today. Imagine the ruckus that cobs created. In 23,000 more years, who knows what corn will look like? If we follow the same trajectory, maybe it will be two feet long and five inches across. Maybe future humans will get to pour moonshine right out of a cob.

Slowly, from the wild bounty of the earth's biodiversity and natural mutations, ancient humans coaxed forth the early stages of crops we have come to know.

Most of our food was developed in seven hotbeds of world plant domestication, where agriculture was independently invented. A man named Nikolai Vavilov figured this out. He was a Russian breeder who created in Leningrad the world's largest collection of plant seeds and earned the Lenin Prize in 1926 for his research on the origins of cultivated plants, as well as his discovery of plant immunity to infectious diseases, a trait attainable by breeding. Vavilov, however, had a dangerous colleague. Stalin's chief of biology, Trofim Lysenko, did not accept Mendelian genetics, particularly the idea that offspring inherit characteristics from their parents. Lysenko followed Jean Baptiste Lamarck's theory of inheritance of acquired characteristics, which meant that external changes in an organism (like losing an ear) would be repeated in the offspring. Lysenko believed that giraffes, for example, evolved from animals with shorter necks by reaching higher and higher into the trees for leaves and stretching their necks. Vavilov criticized Lysenko and thus came in opposition to Stalin. Vavilov was arrested in 1940; the man who had collected seeds of 200,000 plant varieties died in prison three years later of malnutrition.

The Andes was one of the centers of domestication. The other six were China, South Asia, Southwest Asia (the Fertile Crescent), the Mediterranean, Ethiopia, and greater Mexico. Beans, for example, were domesticated first in the Andes of Peru about 5,000 years ago and again in Mexico a couple thousand years later. Native peoples in North and South America also domesticated squash very early on. Seeds of one species (*Cucurbita moschata*) found in coastal Peru have been dated to 10,300 years ago. A species (*C. pepo*) found in Mexico was dated to 10,000 years ago.

Since Vavilov's time, eastern North America, New Guinea, and the Amazon have been added as three more centers of plant domestication. North America contributed to the world table a half dozen or so very minor crops: sunflowers, pecans, blueberries, cranberries. For the most part, as seed saver Will Bonsall would say to me, "Our crops, like us as people, are refugees. Our crops are immigrants."

Out of the richness of evolution and the mystery of life, then, we developed food. It was both a process of discovery and creation, a process that deserves a verb of its own. Maybe creacover. Our ancestors creacovered food in legitimate acts of creation, humans in tandem with nature, no technology required, our fates intertwined.

Over millennia, as humans became dependent on food that we domesticated, the food became dependent on us, a symbiosis of epic proportions. Humans coevolved with food plants like maize and now it cannot exist without us. Human intervention has removed maize's ability to self-perpetuate and it is no longer a "natural" plant. It needs humans. Without humans to care for the maize, the species would soon die out.

And what amazing food we milked from nature. Walk with me through a farmers market, and look at what we can eat. It's a mind-boggling diversity: sugar snap peas, snow peas, carrots, lettuce, arugula, radish, beet, mustard, rhubarb, bok choy, and much more—with multiple varieties of each plant food. In the twenty-first century, we find ourselves—as John Swenson, an elder archeobotanist, put it—sitting at the end of a rainbow. We have pulled our chairs up to a feast, a wealth of plant foods, a groaning board with heaped trenchers.

My son Silas invites me to eat with him in the dining hall at his university and I am rocked by all the choices—pizza or sushi, wrap or grinder, stir-fry or burger. But the real feast I'm talking about is not processed food, boxes and bottles at the supermercado. It's the whole food from which the dishes are concocted: wheat, melons, broccoli, rape, apples, peaches . . . you get the idea.

Behind this food wealth are legions of mostly unnamed and unknown plant breeders across the globe who for millennia shook pollen onto certain corn silks and not others, onto certain stigmas and not others, producing a spectacular fare.

Thanksgiving constantly dwells in my mind.

My saying this may seem crazy when you think about the bounty of the farmers market or the availability of boxes and bottles at the supermarket, but we are, in fact, losing food. Thousands of distinct varieties worldwide, especially ancient breeds, are threatened; fewer and fewer farmers are growing them—and in many cases, no farmers are growing them and varieties are dying out, the seeds for them no longer found. Foods are going extinct. University of Georgia researchers Paul J. Heald and Susannah Chapman searched 2004 US seed catalogs for varieties that had been commercially available a century before. To obtain the names of the vintage varieties, they used the United States Department of Agriculture's comprehensive *American Varieties of Vegetables for the Years 1901 and 1902*, published in 1903. Heald and Chapman found that 94 percent of the 7,262 seed varieties from 1903 were no longer available in 2004 seed catalogs—430 were. This 6 percent survival rate meant a stunning loss of diversity. This study does not even take into account the thousands upon thousands of heirloom varieties never sold commercially.

Surprisingly, Heald and Chapman's study also found that the diversity of varieties available *commercially* actually did not fall much in the century between 1903 and 2004. A total of 7,100 varieties among the same forty-eight crops were listed in 2004, as opposed to 7,262 in 1903. This stasis in commercial numbers is due to varietal replacement—mainly introductions of new varieties, but also imported varieties and heirlooms rescued by preservationists and returned to the market. But here Heald and Chapman were comparing apples and oranges. Open-pollinated varieties that evolved over millennia and whose seed has been saved for generations do not equal scientifically produced varieties. What Heald and Chapman did not analyze are numbers of *open-pollinated* varieties today that were available a century ago. This figure would have more accurately portrayed what we are really losing. A variety lost to seed saving is a variety lost to civilization.

The fact remains that in the last one hundred years, 94 percent of seed varieties available at the turn of the century in America and considered a part of the human commons have been lost.

Three things result from varietal decline. First is the loss to our plates and palates. It's sad to miss, and not know we're missing, all those different kinds of apples, cabbages, corn, tomatoes, and so on. Second is the loss of sovereignty over seeds and the ability to control our food supply.

Third, there's another scary reality to this. All the lost varieties did more than liven up the table and keep farmers independent. Varietal decline threatens agrodiversity. We know this—the less biodiverse any system is, the greater the potential for its collapse. In shriveling the gene pool both through loss of varieties and through the industrial takeover of an evolutionary process, we strip our crops of the ability to adapt to change and we put the entire food supply at risk. The more food varieties we lose, the closer we slide to the tipping point of disaster. We are gazing already into the abyss. Maybe you haven't seen it yet, because maybe you were looking the other way. You were focused on grocery shelves stocked with an overwhelming selection of breakfast cereal.

And the cereal aisle looks pretty diverse, with flakes and puffs and charms and squares. But how diverse is it, really?

Three crops account for 87 percent of all grain production and 43 percent of all food eaten anywhere—wheat, corn, and rice. As Edward O. Wilson writes in *The Future of Life*, these three foods stand between humanity and starvation. "The world's food supply hangs by a slender thread of biodiversity," he wrote. In fact, 90 percent of the plant species recognized as edible come from 103 plants, out of 250,000 plant species known to exist. A small number of plants are feeding a lot of people. Wheat alone meets 23 percent of the world's food energy needs.

When you read the fine print, most of our breakfast cereals contain wheat. I'm walking down an aisle with so many choices my head is spinning and yet it's wheat. We haven't even arrived at the bread aisle.

Let's look more closely at wheat. Traditionally, there were thousands of distinct wheat varieties, as many as 100,000, in fact: Reward, a Canadian heritage wheat; Ethiopian Blue Tinge; Einkorn Gotland, an old Swedish variety; Crimean; Red Fife, a landrace likely of Ukrainian origins; on and on. (A landrace is a local variety of a domesticated species, one with a lot of variety, largely developed through natural means and not through industrial breeding.) Now, half of the wheat grown in the United States is nine varieties. Not only is the breakfast aisle heavy on wheat, it's genetically homogenous.

Crop sameness leads to vulnerability, and not only because it's risky to have everything ripen at once. Ireland's potato blight in 1846 illustrates what can happen in monoculture agriculture; it led to the Great Famine and the emigration of an entire people. About 90 percent of the potatoes eaten by the Irish were a variety called Lumper. It was prolific and an acre could feed an entire family. When a late blight began to wipe out potatoes in Ireland, the Lumper had a slight resistance in the leaves but not in the tubers, which basically rotted in storage.

Think of the consequences if a wheat blight were to descend.

In a way, however, a wheat blight has descended. As Montana farmer and food activist Bob Quinn has said, "Wheat was once known as the staff of life, now we can't even eat it." Modern, industrially bred wheat has been associated with a steep rise in gluten intolerance and with structural changes in wheat proteins that lead to obesity and many other diseases. (Dr. William Davis, author of *Wheat Belly*, has been especially vocal on this subject.)

Not only have our diets become more industrialized, they've become less diverse. Michael Pollan calls corn, soy, and wheat the "building blocks of all processed food." In a talk given at the Georgia Organics annual conference in 2009, he said our diets have changed more in the last one hundred years than in the previous ten thousand. "Monocultures in the field lead to monocultures in food," he said. Diversity of food crops has been dwindling worldwide, and untold numbers of human foods are going extinct. What are at risk are our seeds, especially ancient breeds, and our crop biodiversity. And our health.

Historically, seeds were everyone's responsibility. First they had to be collected. Then they couldn't get wet. Mice and birds couldn't be allowed to trespass against them. Winter was like a river our ancestors had to cross, loaded with waterproof and mice-proof packets and bags.

Furthermore, between growing seasons, the seeds had to be traded, because traditional societies understood that, as in human reproduction, plants do better by outbreeding. To swap seeds is to keep a variety strong and valuable—a genetic currency, the exchange of priceless genetic material. How interesting that the agrarian within us understands that to survive, to keep our food crops viable, we have to be openhanded. Seeds have a built-in requirement for generosity. Otherwise they suffer inbreeding.

Throughout history, if humans up and left a place or a country or a home we had to bring with us the future of food. We have been carrying our food supply with us for at least the past twelve thousand years, kernels tied in leather bags around our necks or sacks stored in large warehouses—a bit like the Japanese Buddhist monks at Daisho-in who've kept a flame lit for twelve hundred years. We've had gigantic winds blowing through the monastery hall, and the fire isn't as big as it used to be, and it isn't so bright. As with most of the biota of earth, we've lost some of the kindling.

— 2 —

a brief history
of industrial agriculture

PEOPLE GET TIRED of carrying crosses. We want to put down our burdens, and the burden of feeding ourselves, as a species, has been our heaviest cross to bear, because if we fail, we starve to death, and because cultivating is full-time, year-round, joint-popping work.

So when somebody came along and said, *I'll do that cultivating for you. I'll save the seeds. You do something else*, most of us jumped at the chance to be free. It was fun and games for a while. While the slaves with hoes, then sharecroppers and mules, then tractors and eight-row cultivators, and now migrants and Lexion 590Rs (80 bushels of corn a minute) toiled in the fields, some of us were back at the house, rocking on the porch swing with a glass of tea, listening through the window to Frankie Laine on the phonograph.

Okay, I'm lying. We were not on the porch. We were not even in the house. We were at the blue-jean factory, double-stitching pocket to leg, pocket to leg, pocket to leg; or on the assembly line, popping screws into automobiles, lawn mowers, and TVs.

Two things happened. First, in that mind-numbing line, when the sight of another motor mount swinging toward us became nauseating, we began to dream ourselves back into the garden and into the fields, where we could hack off the end of a watermelon and reach in, pull out the sweet heart, and lob the rest to the hogs. Second, when Sister boiled up the beans, everybody noticed they didn't taste as good as leather britches, the dish our mamas made using the Cherokee Trail of Tears bean. Or the bunuelos that our grannies made with Eye of the Goat beans. We lusted for the pies our great-grannies made with Pink Banana squash and husk tomatoes.

Wishing ourselves back to the garden wasn't going to change a thing. While we let ourselves be seduced by the dream of ease and plenty, industry was busy salting the fields so we couldn't go back. During what is misnamed—unless you think of the algae from field runoff—the Green Revolution, agribusiness began to woo farmers toward a new dawn. The Green Revolution, which in this country was an entire narrative in itself—a beginning, middle, and end—cemented a slide from agrarian to industrial life. Chemical fertilization, standardization, homogenization, mechanization, and commercialization were catchwords of the new agriculture. It promised an end to world hunger and although it did for a time coincide with the production of more food per acre and worker than ever before, one billion people in the world are still hungry. As Jules Pretty writes in his essay, "Can Ecological Agriculture Feed Nine Billion People?" industrial systems "only look efficient if the harmful side-effects . . . are ignored." Agricultural industrialization rode into town with a skulk of sidekicks: a shift from the local to the global; from the small to the large; from the nutritious to the filling; from the storied to the acultural; from purity to toxification; from independence to victimization.

In order to serve the model of continuous progress, which may not correlate with profits for the farmer, tractors began to take the place of living, breathing workers. My friend Angus Gholson remembers when his father got his first tractor. "I like it okay," the elder Mr. Gholson said, "but nothing comes out the back."

Industrial agriculture also forced specialization, larger areas planted with a smaller number of varieties, as well as more acres planted in annuals and fewer in perennials. Its mania was monoculture, a few dependable, high-yielding varieties that ripen at the same time. "In south Georgia," a farmer told me, "we're trying to major in corn and soybeans." Diversity, in fact, is an *impediment* to efficiency and productivity, as is small or human-scaled agriculture. A diversified small farmer who saves her own seeds is not only of no use to a corporation, but a threat. The basis of diversity—which brings with it possibility, stability, hope, and power—is small and differentiated.

Recently I enjoyed a stint as a visiting writer at Warren Wilson College near Asheville, North Carolina, which has a working farm. The college requires each student to work fifteen hours a week, anything from library support to computer repair to blacksmithing to beekeeping to farming. One day I walked into a classroom to teach, and notes from a previous lecture remained on the marker-board. I was midstream in their erasure when I began to read the phrases and imagine the lecture that had taken place.

> *mechanization of agriculture*
> *gospel of progress*
> *loss of mules*
> *the spectral trace*

I knew the story by heart. The words formed a kind of ghost life I never knew as my own but to which I have longed to return. I have been walking in the spectral trace.

Hybrids

Industrial ag went after seeds themselves and with appalling swiftness took over the seed supply. They began to hybridize, a hybrid being the offspring of a genetic cross. Hybridization is simply plant breeding sped up. The pollen from one plant with desirable characteristics is rubbed on the stigma of another similar plant with desirable characteristics. This flower produces a seed that, when grown, exhibits a combination of desirable characteristics—excellent productivity and growth—called hybrid vigor. Because of this and because new varieties are often bred to be disease resistant, hybrids are tempting to grow.

In the mid-1920s, the first hybrid seeds reached the market in the United States. The first hybrids were two varieties of corn. In 1924, the Connecticut Agricultural Experiment Station introduced Redgreen, and Henry A. Wallace introduced Copper Cross. (Wallace was a corn breeder who would go on to start the Hi-Bred Corn Company, the world's first hybrid seed company, which would come to be owned by DuPont and allied with Syngenta.)

When farmers planted hybrids, bluntly put, they made more money. The bad news about hybrid seeds, however, is that although they often perform better, a farmer can't save them for another year's crop. The offspring fail to "grow true"—to produce fruit similar to the ones from which they came. When a gardener plants seed saved from a hybrid, he doesn't get the same beefsteak tomato or supersweet corn but a hodgepodge of the ancestral strains used in breeding.

Using science to improve food production is not intrinsically bad. Science is worrisome when it only serves the interests of mercenaries and their employees in the long run. Seed companies patent F1 hybrids and have proprietary control over them in an attempt to achieve monopolies on genes. A farmer, then, is forced to return year after year to the company that produced the seed, infecting our food supply with greed.

Increasing numbers of farmers, those that didn't die or quit, went with the new schema—abandoning vintage, traditional, mom-and-pop, place-adapted, planter-bred, hand-saved, carefully guarded seeds. Before 1932, hundreds of corn varieties were grown across the continent, including Stanley corn. Or the grits corn I got from another neighbor, Lewis Snowden, which came from his stepfather, Mr. Gore, who got it from a Mr. Ogden. Or Keener corn, which I can't wait to tell you about.

The year 1932 was the tipping point. This was the year Golden Cross Bantam, the first hybrid maize that became popular, was introduced in US fields and gardens. Stewart's bacterial wilt had plagued farmers; the early 1930s had seen it in epidemic proportions. Golden Cross Bantam, developed by pathologist Glenn Smith of Purdue University, was resistant to the wilt; when put on the market, the seed flew off hardware shelves everywhere.

A sea change happened in the blink of an eye. In 1935, less than 10 percent of Iowa corn was hybrid. Four years later, 90 percent of it was—specifically Golden Cross Bantam. In the slip of time between 1935 and 1939, an interval during which both my parents were born, the face of our agricultural landscape forever changed. Trusting the advertisements, not knowing long-term consequences, not understanding the loss, and wanting to survive, farmers stuck their canisters of homegrown seed-corn on back shelves in sheds and went to town for Golden Cross Bantam. By 1946, according to Jeff L. Bennetzen's *Handbook of Maize*, Iowa was 100 percent hybrid; 90 percent of the corn belt as a whole grew hybrid corn.

Hybrids are designed to be successful in a wide range of climates and growing conditions. They are broadly adapted, as opposed to the more localized open-pollinated varieties—allowing a national and international seed trade to function. Before long, American cornfields transacted almost exclusively in hybrid varieties. To grow them was to enter the milieu of progress.

Nobody faults the farmers, who were acting in their financial best interests and did not know they were joining a system that was already cracked and would be soon broken. Hybridization itself is not really even the issue. As plant pathologist Albert Culbreath told me, "Hybrids have a place and are of use. But they should not be used exclusively and they should be of diverse parentage as well." The real issue is what hybridization represents—including the loss of an extensive seed heritage and agroecological diversity. The problem is the *industrialization* of hybridization.

Suddenly we had a countryside full of farmers who no longer had to worry about leaky barn roofs and varmints in their seed corn. Seed companies worried about that. But the farmers still had to worry. They had to worry about autonomy. They had to worry about parity (as opposed to *disparity*, inequality, or farmers receiving prices for crops that did not reflect the cost of inputs to the farmers). They had to worry about the bank. In giving up seed saving, they became prisoners to Big Ag.

It is important to note that other methods of speeding up traditional plant breeding have come into popularity, including the use of mutagens (agents) to cause mutations in plants. Mutagens include chemicals and radiation (X-, gamma, electromagnetic, ultraviolet). Essentially, seeds are treated with a mutagen and then planted. The offspring are combed for mutations deemed beneficial. These methods roam far from the spontaneous mutations on which ancestral breeding relied.

Genetically Modified Organisms

In the late 1990s came another swing in agriculture, the second speeding bullet striking the very heart of a secure food supply.

Based on the recombinant DNA research of the 1970s, genetically modified (GM) seeds were first planted experimentally in the late 1980s and introduced to American markets in 1996. GM organisms are engineered through their DNA to take on new characteristics. Scientists may turn off active genes, turn on inactive genes, replace one gene with another, or splice in snippets of DNA from entirely different kingdoms of life. Organisms can be genetically engineered for pretty much anything. Whatever the scientist can imagine, she can more or less produce. Bt cotton, for example, contains a bacterium that is shuffled into the chromosome of cotton (there are also Bt corn and potato varieties). Bt, *Bacillus thuringiensis*, is a natural insecticide—a bacteria that produces a spore that proves toxic when ingested by insects. Once Bt is genetically encoded in the cotton, the cotton manufactures its own toxin to kill insect pests. Now we have a plant that not only produces cotton for our T-shirts and jeans, but a bacterium to defy bollworms and perhaps other cotton pests.

Other popular early GM organisms featured a resistance to the herbicide Roundup, the major trade name of glyphosate. They were developed by Monsanto, the company that initially patented and sold Roundup. Farmers would spray their fields with Roundup before planting to kill weeds, because Roundup eventually annihilates every broad-leafed plant it touches. Once the crop had germinated, Roundup was off limits. Now, with the introduction of Roundup-Ready crops— corn, soybean, canola, and alfalfa—farmers can spray anytime, whether the crop is in the field or not. Roundup-Ready plants are in effect wearing raincoats that protect them from the deluge of this chemical. As Kansas corn and soybean farmer Luke Ulrich told National Public Radio reporter Frank Morris in 2010, "There's nothing like Roundup. A monkey could farm with it."

In 2009, a GM corn called SmartStax entered the marketplace. Developed by Monsanto and Dow AgroSciences, this seed reputedly offers eight GM traits stacked in one seed. SmartStax corn produces six insecticidal toxins (to corn borer and corn rootworm) and tolerates two herbicides, glyphosate and glufosinate. These independent traits were not created by repeated genetic blasting, which requires the insertion of DNA sequences, but by crossing existing transgenic corn lines.

Again, farmers embraced GM corn in the same fashion they had taken to hybrid corn. In just over a decade, over half of all corn grown in the United States was GM. You may be wondering what's so wrong with this. If science can work to our advantage, let it. If wonderseeds can feed the world, let them. But there's plenty wrong.

First, the genetic insertions are "cheater" genes—a farmer can't see them, can't prepare for them, and can't protect a farm from them. Second, why would it be important for a chemical company to develop a product such as Roundup-Ready soybeans? To minister to the hungry? To protect our environment? To serve human civilization? Or to sell more chemicals?

Second, for the first time in the long and marvelous history of humankind, genomes can be owned. Companies now patent varieties of plants, especially the new scientifically produced, advanced cultivars. The genomes of wild rice, the only grain indigenous to North America and a vital staple for the Ojibwe people, were patented by a California company in the late 1990s. Subsequent genetic tinkering by scientists at the University of Minnesota that produced several new strains outraged author and indigenous activist Winona LaDuke. "We have a 2,000-year-old relationship with wild rice," she said. "Conceptually, it seems almost impossible to patent something called *wild*."

Read closely here.

Some things are inherent to the earth and thus belong democratically to all its inhabitants. Air and water, for example, are part of the public domain and should be forbidden in the marketplace. Seeds—always part of the great commons of human history—can no more be owned than fire. Or the ocean. And yet, the biotechnology industry has steadily made its way through courts and legislative halls like an evil maggot, claiming what does not belong to it, saying life can be owned. And it can't, Monsanto. It can't, Syngenta.

A seed recipe is not real property, of course, but intellectual property—a legal invention, meaning an idea accepted or designed by courts and then used to uphold certain interests. The idea of intellectual property rights is a legal invention because life belongs to all of us.

In addition, seeds not already registered as a known variety may be snatched up and patented as intellectual property by anyone. The multinationals are particularly effective at this. For example, Monsanto patented an Indian wheat used for chapati for eons. In another ludicrous case, a Colorado man patented a variety of yellow bean that was ancient in Mexico, then demanded royalties from Mexican peasants.

Another major concern with genetically engineered organisms is that vectors are necessary to insert chosen genetic qualities into plants, and viruses and bacteria—including some proven harmful to humans, such as *E. coli*—are used for these vectors. Genetic tinkering is not something you can see. An ear of GM corn looks about the same as an ear of non-GM corn. You can't see that it contains a virus. Because most GM products are released in the United States without independent environmental or health testing, nobody knows exactly what effects these organisms will have on humans. The approval and release of

GM foods into the United States is a huge, unplanned, untested, unpredictable experiment—and we eaters are the lab mice.

The most egregious part of the story is that corporations own the playing field. They control the government regulators, or more aptly, they *are* the government regulators in a revolving-door parade between the multinationals and government. We could call it cross-pollination, but I won't use a metaphor of life for such a destructive practice. Michael Taylor, to name one of dozens of examples, was a Monsanto executive before he became the FDA's deputy commissioner of foods. Genetic engineering is not evolution sped up; it is evolution in the hands of multinationals. Thanks to hybridization, genetic modification, and seed patents, and with government as the enforcer, a handful of individuals have control over what we eat.

Loss of Biodiversity

In both hybridization and genetic engineering, farmers lose control of the ability to save seeds year after year and to breed plant varieties ideally suited to a place. To hell with natural selection. Gene flow be damned.

Genetic freedom allows crops to adapt and evolve—to disease, to a transforming earth, and to macro- and micro-climates. These adaptations often prove useful and profitable. A landrace of alfalfa, for example, discovered in 1940 in Iran, proved resistant to stem nematodes, and this landrace parented a strain of alfalfa that greatly improved production. Biotechnology subverts natural adaptation and destroys diversity.

In the United States, according to the United States Department of Agriculture (USDA), this is gone: 95 percent of vintage cabbage varieties, 96 percent of field corns, 94 percent of peas, and 81 percent of tomatoes. In the century between 1804 and 1904, over seven thousand varieties of apples were grown in the United States (including Roxbury Russet, Black Gillifeather, and Greening)—and 86 percent have been lost. Sure, we've kept lots of varieties and we've gained new varieties, but 86 percent have disappeared.

Ever fewer commercial varieties in chemical systems replace local varieties employed in traditional farming systems. In 1949 China, ten thousand distinct wheat varieties could be found growing. By the 1970s, those varieties had been laid to waste—one thousand were in use. Likewise, in Korea, by 1993 only 26 percent of those garden crops growing in 1985 were still being grown, a loss of three-fourths in eight years.

Loss of Landedness

Pull one thread on a problem, like that of the loss of food diversity, and you'll find one issue connected to everything else. Part of our loss is that we've not stayed in one place. For the past century, rural places have steadily bled people. The sons and daughters of farmers were lured to cities, resulting in the largest diaspora in American history, one that spanned many generations and continues to this day.

In 1900, 41 percent of Americans farmed for a living. Now less than 2 percent earn their livelihood farming, according to the USDA's Economic Research Service. During this era, many folk traditions went underground—not just seed saving, but customs like buckdancing, clogging, fiddling, tatting, and quilting.

The falling apart of rural communities began in the late 1800s with the advent of railroads that made travel easier and brought a keen dissatisfaction with rural isolation and lack of society. The bleeding of rural people intensified during World War II when, to rebuild our war-broken country, the US government launched an advertising campaign to entice people away from farms to cities. Industrial capitalism needed a workforce, and what it promised in return was certain prosperity. Jobs were plentiful in the city, factory labor easier than hardscrabble farm life. To leave the farm was an act of patriotism (albeit a misguided one).

The ad campaign worked. There ensued a mass exodus of rural people. Between 1915 and 1960 about 9 million rural Southerners, to name one region, were displaced to cities; another 9 million—approximately half white, half black—were gone from the South entirely. (We must also recognize that to leave could also be an act of self-preservation. In the case of the South, millions of blacks fled rural lands to escape Jim Crow laws and search for higher paying jobs and freedom from cultural oppression.)

Young people would graduate from schools in little towns like Ideal, Georgia; Liberty, Mississippi; Enterprise, West Virginia; Faith, South Dakota; and Hope, Arkansas—and they would go away to university and never come back. They would have internalized what we told them, as the late thinker Paul Gruchow explored in his book *Grass Roots: The Universe of Home*, that if they wanted to amount to anything they'd better leave home: "We raise our most capable rural children from the beginning to expect that as soon as possible they will leave and that if they are at all successful, they will never return. We impose upon them, in effect, a kind of homelessness." If they were any good they wouldn't be in the country—they'd be somewhere else. They would want to pursue learned professions. They wouldn't want to tackle illiteracy, religious fundamentalism, poverty, joblessness, racism, rural homophobia. People speak of having "escaped." "I would die if I had to go back there," I've heard said. "I couldn't wait to leave. Nothing's there." Once uprooted, folks tend to continue peregrinating, moving about for careers, for education, for marriage, for lifestyle.

Four-fifths of people in the United States now live in urban areas. Across the country you see evidence of this "hollowing out" of rural America—abandoned small farms, ghost towns, country stores with dark windows—and its attendant suffering. Rural places have hemorrhaged their best and brightest children, their intellectuals, thinkers, organizers, leaders, and artists—those who would create change and who would parent another generation of thinkers. All gone.

Our seeds are disappearing.

When seed varieties vanish from the marketplace, they evaporate not only from collective memory but also from the evolutionary story of the earth. Seeds are more like Bengal tigers than vinyl records, which can simply be remanufactured. Once gone, seeds cannot be resurrected.

Goodbye, cool seeds. Goodbye, history of civilization. Goodbye, food.

A seed makes itself. A seed doesn't need a geneticist or hybridist or publicist or matchmaker. But it needs help. Sometimes it needs a moth or a wasp or a gust of wind. Sometimes it needs a farm and it needs a farmer. It needs a garden and a gardener.

It needs you.

me growing up

WHEN I GREW UP in the 1960s and 1970s among south Georgia's flat gray fields, which spun waist-high tobacco and battalions of cornstalks out of thin dirt, our culture was agrarian. Seeds were currency, seeds were gifts, seeds were steaming bowls of the future.

My grandmother Beulah held stock in this economy. She loved plants and all that they produced. She had an ingenious and fascinating greenhouse—a pit house, dug into a red clay bank beside the dirt road. From the outside, the house looked about four feet high and was covered in plastic sheeting. When you pulled open its little door, concrete block steps led three feet down to a small earth room with a dirt floor and damp clay walls, ferny and mossy, a basement without a house. The pit house was steamy and hot and smelled like geraniums. The air was algaeic, it was so green.

We grandchildren had to ask permission to go into the pit house. A child had to move carefully there. She could not knock off blooms or break stems. She had to watch for snakes. She had to care about what she was seeing.

In a greenhouse, transformation is possible, because the lead tendril of the Willow-leaf Running Butter bean searches for your soul. A garden is the same. The Sunrise Serenade morning glory vine binds you. Maybe transformation is possible anywhere there are profusions of plants.

After my grandmother was too old to care for it, the greenhouse was dismantled and the pit filled with clay. But not yet. For a while I could learn my grandmother's wisdom. I could observe her movements among African violets, her hands among links of Thanksgiving cactus. I could watch the clear water gushing from the hose.

My heart first opened to gardens in the time of cowpeas. I remember one specific day. It happened before my mother stopped wearing pants, before she adopted a religion that forbade them. On this day she wore dark blue pedal pushers because she had come to her mother's garden, since she herself did not have one; she had a junkyard, what she and my father called a wrecking yard. She had come to help my grandmother put up peas—field peas, not garden peas.

My skinny, gentle sister was wearing pants with little flowers printed on them and flat-bottom sneakers of light blue. She was six, holding my hand. I was two. My mother was picking, hoping my baby brother back at the house with Grandmama would stay asleep long enough for her to fill her pails. With this many children, three so far, how to even find time to put food by? My uncle was also in the pea-patch with my mother, helping harvest the peas that Mama and Grandmama would shell back on the porch. My sister walked me toward the edge of the garden. The peas had grown out of their lines and tangled together. They made an undulating sea that threatened to wash over us.

"Sis, watch out for snakes," my uncle called to my sister. All the females in my mother's family get called "Sis." "Maybe y'all best go back to the house. I'd hate to see a snake bite you." The vines smelled like thunder.

I didn't think there were snakes beneath the big leaves. I stepped into them. I tripped and fell. My uncle looked up but didn't say anything. I got up, brushed my palms together. My sister grasped my hand again. She listened better than I did. I pulled away. I reached for a long thin fruit. I think it was born in me to eat the fruit—I knew not to eat the leaf or the vine.

My mother was picking, picking. She couldn't pick one lime-green bean pod and sit down among the upright leaves and begin to chew on it, as I could. Its taste was strange. I arrived at a moist round pea. Its taste was stranger. I chewed it with my new teeth.

"Mama," my sister yelled. "She's eating the peas."

My mother turned her heart-shaped face toward me. It was often turned elsewhere. "You're going to get a bellyache," she said.

I was glad she didn't tell us to go back to the house. I looked at her and smiled and kept chewing. I was so fat my face unfolded when

I smiled. I didn't know how to say what I was feeling. The peas did not need to be cooked.

"I don't guess they'll hurt her," my mother said to herself as much as to my uncle or my sister. She was filling an enamel milk bucket. We would spend the afternoon taking these peas from their shells and tossing the empty pods into bushel baskets. I would try to play in the pods.

I knew my mother was tired. I knew she needed more sleep. But she was young, she was still in love with my father. She would always be in love with him. She had found her calling. I stood unsteadily among the tripping vines, clutching the fat pods with my fat hands.

"Peas," my mother said. "Can you say that?" I looked at her figure, tall and thin against a blinding sun. I squinted, then grinned the fat grin.

"Pees," I said.

The only gardening mentor I had was not saving seed because she understood genetic erosion. She was saving seed because she had learned how to do it when she was young, because she had always done it, and because it was the natural thing to do. She moved about her kitchen with her graying hair clipped short, baking lemon cakes and filling her pantry with pear chutney.

I was young in her kitchen when I saw my grandmother scraping squash pulp to dry on paper napkins smoothed on a foil pie plate.

"What are those, Grandmama?" I asked, making my voice small. I knew to be polite. I knew to ask only crucial questions, those to make adults feel important.

"Cushaw," she said, proudly.

"Cushaw." A funny word. Like a sneeze. "What is that?"

"Oh, a kind of pumpkin."

"Do you eat the seeds?"

"No, child, you plant them," she said. "To grow more cushaws." She looked at me with a mixture of love and pity. "We'll dry them and plant them in the garden next year and more pumpkins will grow."

"Oh."

After supper, I knew, my grandmother would go out to rock and watch the hummingbirds in the impatiens lining the screened porch, and she would let me climb up onto her lap. My grandmother had a soft bosom. Her housedress felt electric against my cheek. She smelled like the talcum powder in the round container on her dresser.

"Grandmama, will you show me how to grow cushaws?" I asked.

She laughed. "I will," she said. "But I don't know how much growing you'll be able to do, with you starting school this year."

I am immensely grateful for everything Beulah taught me. Without her, where would I be? "The branches grow because of the tree," says a native Hawaiian proverb. We are here because of our ancestors.

Later, a ubiquitous first-grade experiment in germination—bean-growing in a milk carton on a classroom windowsill—affected me deeply. A kid has no power, everybody knows that, and yet when I planted a hard little round rock in a handful of soil, after a week or two something started to wend up out of the dirt. Two pages of a green book opened and rose skyward on one slender green leg. Days passed, more leaves opened, and the thing kept growing. I wanted to run and shout the news. Wizardry had happened, all because of me.

How could something so small shape-shift into something so big? Because seeds contain more than they are, life's design held in a speck of germplasm. They are the source of all life, miracles in tiny packages. Magic. Just add water.

I am crazy about seeds in part because they are metaphors for so much: innovation, potential, multiplication, plenty, the future. I am also crazy about the literalness of seeds, the bundles of energy that have propagated plant life since flowering plants first appeared on the earth.

As a child who grew up isolated, without television, I sank deeply into stories. My father, a junkman, brought home loads of miscellany that sometimes included boxes of books. In one such outfit arrived a discarded collection of stories, a literature textbook for elementary school students. I read this book over and over. In the stories, someone's very life was always at stake (just as it is in real life, now). A servant would be beheaded if he didn't invent a new dessert to please a young spoiled prince, who wanted something both hot and cold (voilà! the ice-cream sundae). In another, a boy tried to take off his hat to honor the king, except a new hat kept replacing the one he took off. The hats became more and more elaborate. The boy was climbing the stairs of the death tower when the thousandth hat, the last and most magnificent one of all, appeared on his head. When the boy removed this crowning glory, finally his head was bare, and now he could bow down properly to the king. The king was so taken with the bejeweled, beplumed hat that he forgave the boy's impertinence, spared his life, and made him a prince.

My favorite story was of a pioneer family who lived in a log cabin in the West, eking out crops. One evening, looking in his mother's sewing basket for a button, the hero of the story found a single seed. Excited, he showed his family, who encouraged him to plant it. When spring arrived, he did exactly that; the seed germinated, and it grew into a pumpkin vine. The boy's family was able to eat pumpkin pie at Thanksgiving.

I have never forgotten the story of the wealth in a single seed.

Some years later my grandmother gave me the first heirloom seeds I ever began to save, although neither I nor she knew the terms *heirloom* or *vintage* at the time. I was spending Saturday with Grandmama. By the time my sister, brothers, and I had trooped through the front door that morning, she had picked a hamper of Zipper Cream peas and we spent the morning helping to shell them. At lunch I noticed a handful of strange seeds in a cut-glass bowl on a buffet in the dining room.

I waited until Grandmama finished cooking. I waited until after lunch. I waited until my sister and I washed all the dishes in the low sink, the one built for a woman who was barely five feet tall.

"Grandmama, what are these?" I asked. The beans were large as eyeballs; they looked like eyes, white with dark irises.

"Sis, Ermalou gave me those. She called them Jack beans."

"Can you eat them?"

"Ermalou said she thought you couldn't. I don't think I'd attempt it."

At home, I'd claimed a section of a flower bed for my own garden. I wanted one of these eyeball seeds to plant in it. To steal one would be a sin. I needed to word my questions carefully.

"Are you going to plant them?"

"I did already."

"Are these extra?"

"You might say that."

"I'd like to try planting one."

"Well, honey, help yourself."

That was all I needed. The seeds felt like oblong marbles in my hand. "Why are they called Jack beans?" I asked.

"After Jack and the beanstalk, I reckon."

"They grow very high?"

Grandmama laughed, "You'll have to plant them and see."

"Thank you, Grandmama." Then I asked her if she knew what was orange and half a mile tall.

She didn't.

"The Empire State Carrot," I said.

She laughed. "I hope you get a Jack bean that high."

"The Empire State Bean."

Within a couple of years I'd set a brush fire clearing land for my second garden. It was a patch of ground about as big as an American bathroom at the periphery of the junkyard. I was twelve and went around talking about George Washington Carver, one of my heroes—a man who healed sick plants, painted with clay and pokeberry, and invented peanut butter. His influence would lead me to find amity and lasting friendship in the company of botanists and seedspeople.

Burning the weeds would be easier than digging them out, I thought. I found a pack of matches and touched a small flame to one corner of the patch. The fire leapt across dry broomsage, out of my plot, underneath a defunct moving truck in our backyard, and into the junkyard. I had no water to fight it, only a rake. My yells roused my family, and my mother and brothers came running. The only damage was that for a few weeks I suffered repeated lectures on fire safety, complete with narratives of explosions, burn victims, and homeless families. I had to clear the little garden by hand.

No one in my nuclear family gardened, so I sought and received guidance from my grandparents. The Jack beans grew terrifically skyward, producing large, floppy leaves and foot-long pods that dried and rattled and, when I shelled them out, amazed everybody. Okra poked its pods into my hands. The yellow squash bore fruit. I came to feel almost drunk with gardening.

I got crazy about seeds because I was crazy about plants because long ago I realized that the safest place I could be was in the plant kingdom—where things made sense, where the malice we have to contend with in the animal world was absent, where nothing was going to eat you, really.

I remember the relationship I as a child had with plants. They were the things that came closest to my body, so that I was intimate with them—with trees, with wax myrtle, with daylilies—in a way that I was not intimate with anything else. The transplanting of Johnny-jump-ups or the picking of roses was extremely personal. In place of boyfriends, I had honeysuckle from the vine, radishes from the ground, asters from the ditch. To touch something is to develop a relationship that is sensory, one that is personal and thus private. This is not the language of botany, but of friendship. I made friends with the flora.

One spring, in science class, I had to construct a poster of seeds, and so I went willy-nilly about the house and junkyard—a little Bartram, full of ingenuity and enthusiasm, collecting. Wasn't even a grain of rice a seed? I glued corn kernels and red fleshy magnolia seeds and orange pits and watermelon spits to my poster board. My display was heavy with acorns, airy with white wisps of dog fennel. It earned an A.

An entrepreneurial child, I wrote to the American Seed Company in Lancaster, Pennsylvania: "Please send me your BIG Prize Book and one order of 45 packets of American Seeds. I will sell them at 40¢ a pack, send you the money, and choose my prize, or I may keep one-third of the money instead of a prize."

Every year, through the mesmerizing and tumultuous narrative of my childhood, I planted, I watched, I learned, and sometimes I harvested. I was still sowing that garden the year I was a senior in high school, newly eighteen, when my friends were dating and drinking and going to movies, and I was most definitely not, because my parents strictly did not allow such carrying-on.

The nature of my life was apparent in a journal entry dated May 14, 1980, a Wednesday only a few weeks before I graduated from high school. I was eighteen years old. The boys to whom I refer were my two brothers, one and two years younger, both of whom could drive long before I. I had a job after school shelving books and filing articles at the public library.

The boys picked me up from work this afternoon, and we went out to Thompson Farm Supply. I love the atmosphere of feed and seed stores. I got a pound of peanuts for planting ($1), a scoop of squash (20¢), and a big packet of turnip seed ($1). At the grocery store the other day I had bought some of those colorful packets of seed: okra, radish, marigold, and dwarf zinnia, 49¢ each. I sowed the grainy turnip seed in a three-foot bed with a rake. I made rows and planted squash and okra. Then I hoed out a few weeds. The wind came up, it got cooler, but the clouds didn't collect. Finally around nine (Mama and I were jogging) it began to sprinkle. Rain has fallen lightly since. Grow, grow, grow!

Eighteen years old! What was I thinking? I sound like a ten-year-old. Why wasn't I sneaking out my window at night to join friends driving back and forth between the Dairy Queen and the Methodist Church? Soon enough I was out in the world, seeing it for myself, as much of it as I could see. In a world where there was so much to love, I came to love plants and, accordingly, seeds.

— 4 —

sycamore

I BEGAN GARDENING SERIOUSLY when I was twenty-one. In the early 1980s, I used scholarship money from college to buy land, and as a college senior I moved to this twelve-acre homestead in rural north Florida, west of Tallahassee. Already a hippie community was in full swing there in a place called Sycamore, not far from Greensboro, the next biggest town being Quincy. My place, Hoedown Organic Farm, was at the dead end of an unnamed dirt road.

In this little community of back-to-the-landers dwelled a number of gardeners who taught me what they could and inspired me to carry plant-love to new heights. All of their gardens were organic and magnificent. Sara introduced me to comfrey and chayote and luffa. Lesa grew foxglove and brought me starts of bee balm and lavender. The Fishers, macrobiotic neighbors who had owned a nursery in south Florida, had planted an incredible orchard and grew daikon radishes and Japanese greens. Once or twice a year somebody hosted a plant exchange.

In 1984, twenty-two years old and high on gardening, I ordered a strange variety of squash from the *Market Bulletin*, a weekly publication of the Georgia Department of Agriculture, which was full of free ads. Candy Roaster, the farmer had called the squash, and described it in his ad as nothing you've ever seen in a feed store or a seed catalog. The squash grew two to three feet long and over six inches in diameter, like a stout, curving club. It was dark pinkish orange when ripe and scrumptiously sweet.

When I ordered it, Candy Roaster was simply a novelty to me. About that time, however, as a young granola in Sycamore, I read about the Seed Savers Exchange. It was an emerging group trying to preserve heirloom seed, mostly through the exchange of seeds by members. The Seed Savers Exchange had begun a decade earlier, in 1975, thanks to Diane Ott Whealy and her then-husband Kent Whealy. The couple was taking care of Diane's ill grandfather, Baptist John Ott, who had been growing seeds brought by his parents from Bavaria when they immigrated to St. Lucas, Iowa, in the 1870s. One was a blue morning glory with a magenta center and purple rays, which the Whealys called Grandpa Ott's morning glory. The other was a German Pink tomato. When Grandpa Ott died, the Whealys realized that only they were left to keep their family heirlooms alive, a fact that introduced them to the knowledge that everywhere, old varieties were dying out. Many traditional varieties of vegetables and flowers, planted and saved year after year by family farmers and gardeners, were being lost or only grown by very few people—and sometimes only one person. They determined to keep seeds alive. Their exchange began a movement of gardeners who stalked rural America, questing for heirlooms.

I decided to join.

I had a new interest now in the Candy Roaster. I sent a letter to the *Market Bulletin*, inquiring if this was a new species, an odd squash invention, or an antique? Might there be other old varieties like this that people had been keeping alive?

I found in the paper where you wanted to know about those candy roasters. I have plenty of those seeds. I raised 13 good ones this year. They sure make a nice pie. I thought I would let you know they are still around.

Bertha Woody
Ellijay, Georgia

P.S. I have a son living in Orlando, Fla.

I'm so glad your Canney Roaster Sweet Potato pumpkins have done so well. And indeed I think they are delicious also. Some people pick very young and fry like yellow squash but I let mine get big and dry and cook like pumpkin. I've only had them a few years maybe 6 or 7, but Granny Dills is 86 yrs old and she is who told me they are Canney Roasters and she said she ate them in her younger days so I do not know how old they are. I have never seen it in a catalog but I've never looked much either. I got seed from my husband's first cousin and he called them tater pumpkins. An older relative down the road has relatives in N.C. and he said they raise them up there but they also call them Canney Roasters.

Sincerely,
E. Wise
Dahlonega, Georgia

I saw your letter in the Market Bulletin. I was born in Luthersville. My daddy still lives at Grantville, Ga. It's down below Newnan. I love all kinds of odd flowers and vegetables. I don't have a lot of the old-timey vegetables that you wanted. Would send them if I did. I have the peter pepper (hot) and cow horn pepper. I wondered if I sent you the money would you send me a banana tree? Or tell me of a place down there where I could order things like that. I've got a pineapple that I started from a pineapple. Also an avacado. Thanks so much.

Love,
Frances Campbell
Paducah, Kentucky

P.S. I was born July 21, 1933.

A friend, Irwin, and I nailed together a one-room, off-the-grid, tarpaper shack in Sycamore. I'll be generous and call the structure, which was constructed of heart pine planks and two-by-fours recycled from defunct tobacco barns, a cabin. Irwin and I weren't engineers, much less carpenters. We were youth scoffing at a capitalist society.

Even before the cabin was finished I had a garden started. I planted only open-pollinated varieties, since I wanted to save my own seeds and keep food's gene pool strong. I was ordering seed from small companies like Johnny's Selected Seeds and Pinetree Garden Seeds, which marketed inexpensive, family-size packets for the home gardener. (In 1985, sixteen packets cost me $6.85, including shipping.)

I was also ordering from seed savers. The packets poured in: Granny's Scarlet Runner bean, Haitian green, New Mexico Cave bean, Genuine Georgia Rattlesnake watermelon, Calico Crowder cowpea, Millhouse Butter bean, Chocolate Sweet pepper, Old Sugar gourd, Self-seeding lettuce, Byrd mustard, Cinnamon vine, Ada soybean, Old Timey melon.

Bulgarian Triumph tomato, Red Sausage tomato, Czar tomato, Truck Gardeners Delight tomato, Geisha tomato, Peron Sprayless tomato, Red Currant tomato, Mule Team tomato, Florida Pink tomato, Climbing tomato, Manasota Volunteer tomato, Super Italian Paste tomato, Dinner Plate tomato, German Pink tomato, Old Handed-down Pink tomato, Arkansas Traveler tomato, Mr. Charlie tomato, Old Brooks tomato, Czech's Excellent tomato, Florida Basket tomato, Believe It Or Not tomato, Moneymaker tomato, Deweese Streaked tomato, Stone tomato, Firesteel tomato, Yellow Cherry tomato.

Red White and Blue Indian corn, Bachelor Button, Squaw bean, Black Becky bean, Hornet's Nest gourd, Rice pea, Hopper's Flower Garden okra, Blacklee watermelon, Aconcagua Sweet pepper, Blue flax, Wahirio tobacco, Listada de Gandia eggplant, Garden huckleberry, Cowhorn turnip.

The garden journal I used that year was a red, hardback appointment book from 1980, a date I scratched through and changed to 1985. On March 9—a Saturday, not a Sunday—I transplanted Bibb and Tom Thumb lettuce. On March 24, I planted Golden Midget sweet corn near the clothesline and two varieties of watermelon, Crimson Sweet and Congo, in mounds filled with fish heads hauled in barrels from a fish house. On May 9, home all day, I planted Blue Lake pole beans and Tahitian squash, as well as nicotiana, eggplant, zucchini, cleome, chamomile, and jalapeño pepper.

We lived in Sycamore without running water and caught rainwater in buckets and barrels off the eaves of the tin roof—call it walking water—which we used to water the garden. We bathed in the stream that ran along the northern edge of the property and hauled in potable water.

That spring came a drought. On May 20, I wrote, "The sun is harsh, sending waves of fire, sucking water from the earth, giving snakes power to strike." The rain buckets and barrels caught a few pathetic drops of dawn dew that evaporated before midmorning. Obviously, by June 4, 1985, I knew something about seed saving: "I hope I develop drought-resistant, heat-tolerant strains of vegetables."

When rains finally came, the gardens at Sycamore grew divine. They were living art, a verdant jumble. I had a sun garden near the cabin, a circular mound with six beds radiating. I had a pomegranate garden with a bench, made with two rocks and a plank. Pumpkins and melons, including a hand-sized Japanese melon whose skin was edible and which I have not been able to find since, sprawled among wild persimmon trees. There was lettuce leaf basil and holy basil and sweet basil. Kale and chard grew lush in long raised beds.

Anything strange and unusual, I tried. Unicorn plant and castor mole bean. Scarlet runner bean. Green cotton. Myrrh, jicama, and alyssum never germinated, but the rest bounded for the sky. Sometimes I entertained myself with a thought experiment: If I were given an acre of bare soil on a far island and I could bring one plant for comfort and joy, not to sustain me calorically but to enjoy, what would I take with me? I might choose belladonna, for each bloom would be a trip not taken; or moonflower, glorious evening delight. I might choose marijuana, for the filigree of its odd-green leaves; or passion vine, its flower the complicated and intricate formula for so many stories—the twelve disciples, the twelve wise women, the dozen eggs. Oh, how could I choose? I have never seen a plant I did not love.

In May, I reported having found sprouted date pits in the compost bin. Later that summer, on August 23, I gathered seed from mullein, four o'clocks, and nicotiana.

A few negative metaphors are associated with seed saving. For a vegetable to flower has been considered by gardeners as a mistake—

oops, it went to seed, yank it out! Going to seed has meant that a person has gone wayward, and seedy places are unsavory. A seed, however, finds its nativity in a flower, a thing of beauty, color, fragrance, form, and variety. Flowers are food for the soul. And the seeds they fashion are life, sustenance, the future. We are utterly dependent on them. Seeds are the bridge between us and the sun, emissaries of the solar system, bundles of cosmic energy.

— 5 —

what is broken

BEFORE WE GO FURTHER, I want to make sure you understand what is broken.

When the United States invaded Iraq in the spring of 2003, the plan of attack was strategic. Our government showed no concern for Iraq's cultural resources, including its seeds.

Iraq's history is one of seven thousand years of civilization. Located in the Fertile Crescent, an arable oasis considered by scholars to be the cradle of civilization, Iraq's roots date to Mesopotamia, which flourished on the banks of the Tigris and Euphrates Rivers. The region is credited with producing the world's first writing, first calendar, first library, first city, and first democracy. "The US government could not have chosen a more inappropriate land," said novelist and activist Arundhati Roy in her acceptance speech for the 2002 Lannan Prize for Cultural Freedom, "in which to stage its illegal war and display its grotesque disregard for justice of any kind."

By 2004, the United States put in place a new foundation of governance for the conquered Iraq: one hundred orders enacted by Paul Bremer, chief of the Coalition Provisional Authority. One of them was particularly strange. Under the heading "Amendment to Patents, Industrial Design, Undisclosed Information, Integrated Circuits and Plant Variety Law," Order 81 authorized the introduction of GM crops and instituted intellectual property rights for seed developers. The order made seed saving of GM varieties illegal and forced farmers who used GM varieties to purchase seed each year.

Order 81 was not a law adopted by a sovereign country. This law was not enacted out of distress over the nourishment of Iraq's people. The law's lone purpose was to open a new potentially lucrative seed market for the multinationals that already controlled seed trade in other parts of the world. The corporate giants had a new market for their wares.

At the time of the invasion, five million acres of wheat were under cultivation in Iraq. About 97 percent of Iraqi farmers either saved seeds or purchased locally grown seed from nearby vendors. Untold varieties were being grown, including Saberbeq, a wheat landrace grown in northern Iraq that was known for the quality of its bread and its drought tolerance.

"Farmers shall be prohibited from reusing seeds of protected varieties." At first blush, the language of Order 81 seems benign enough. Restricting seed saving of protected varieties did not interfere with traditional practices, did it?

It did, for two reasons. First, behind the backs of the farmers, the sneaky multinationals were quietly patenting seeds that indigenous farmers had manually and painstakingly developed through selection and over time. Suddenly, then, a farmer might not own a seed he had been growing and saving. All a corporation needed to do was hold a few farmers accountable for thievery through threats or outright lawsuits and the rest would cave.

Second, US Army soldiers, in a program dubbed Operation Amber Waves, were handing out free wheat and barley seed, free fertilizer, and free T-shirts emblazoned UNITED FARMERS OF IRAQ in several districts of the country. The seeds were GM. They ensured the genetic contamination of Iraq's heritage crops and foretold the liability that would come to bear on Iraq's growers. Call them seeds of empire.

"Introducing transgenic wheat means replacing this diversity and leaving it to extinction," said Nagrib Nassar, professor of genetics at Universidade de Brasilia. "It will be replaced by a monoculture with a very narrow genetic base. This is a problem. This will be a catastrophe."

In a separate advance in this war on farmers, consider this—Abu Ghraib, a town outside Baghdad, was conquered and controlled by invading troops. The country's gene bank was located at Abu Ghraib, where Iraq's germplasm securities were looted and lost. Sanaa Abdul Wahab Al-Sheikh worked at the national gene bank at Abu Ghraib. During the invasion she hid about a thousand accessions, items in the collection, both underground in her backyard and in her home refrigerator, and was able to save them. She now works at the rebuilt facility and is traveling the country adding seeds to the gene bank. But ancient landraces of grains are disappearing as Iraqi farmers grow new GM varieties. The foundation of Iraq's food sovereignty eroded in a flash flood.

Jeremiah Gettle of Baker Creek Heirloom Seeds offers for sale the seed of the Ali Baba watermelon, an Iraqi fruit collected in the 1990s by Aziz Nail. As Gettle says in his description of the melon: "It is now nearly impossible to get seeds from this ancient country whose people have lost much of their genetic heritage in the long, bloody war. Now our corporate agriculture has been kindly suggested to native farmers who are losing thousands of years of plant breeding work; I guess they have gained the freedom to sign a patent waiver and support our genetically engineered greed." Baker Creek also offers four Iraqi tomatoes: Basrawya, Ninevah, Al-Kuffa, and Tartar of Mongolistan.

What happened in Iraq was part of an extremely well-funded, politically savvy, militarily connected, corporate-minded gambit to control the world's food supply. That's your food, too.

How the American Food System Is Broken

1. OUR FOOD IS GOING EXTINCT.
In the last century, 94 percent of vintage, open-pollinated fruit and vegetable varieties vanished. By 2005, the United Nations reported, 75 percent of the world's garden vegetables had been lost. We're hemorrhaging old varieties, despite the productivity, adaptation, and delicious taste of many heirloom strains.

2. OUR FOOD SUPPLY IS BEING STOLEN FROM US.
I blame the shrinking of our food choices on corporations. I almost never blame anything—save perhaps a willingness to believe lies and be cheated—on people. At the expense of our soil, our health, our culinary traditions, our environment, and our communities, traditional ways of growing food have been stolen by corporations and replaced with chemicals and radicalized seed.

Two concepts are important to this conversation. One is the idea of vertical versus horizontal production. Vertical production of food means control from the top down. Horizontal production means the control is at the level of the producer or eater. When I bike past a cornfield with a little sign that says P1376XR, which is a DuPont GM corn, I feel crushed. But when I think about the organic or biodynamic or naturally certified farmers who sell at our farmers market, on their little farms toiling in their human-scale fields, I stand shoulder to shoulder with them.

In my mind I see a map of our coastal plains region, stretching from the geologic fall line toward the coast. Within a morning's bike ride I can reach Debra and Del Ferguson in their grassy pastures (Hunter Cattle Company), Relinda Walker with her acres of purple and orange carrots and canary melons (Walker Farms), Jimmy and Connie Hayes with organic peanuts (Healthy Hollow Farm), and Cindy and Larry Kopczak with their pecan orchard (Snug Hill Farm). I could put on my boots, start walking, and get to them.

On the contrary, I could walk for years and never find Monsanto in his field or standing in front of his tailgate at the Saturday market. Monsanto is in the sky.

The second concept is the idea of one large entity versus many smaller ones. Farmers have been told for three decades now: get big or get out. The new adage, in direct opposition to the industrialization of agriculture, is *get small or get out.*

3. OUR FOOD SUPPLY IS BEING BOUGHT OUT FROM UNDER US.

The people will eat what the corporations decide for them to eat.
—Wendell Berry

In the 1980s and 1990s, chemical companies aggressively moved to purchase privately held seed companies in order to capitalize on the profits that genetic engineering promised. The modus operandi was to buy a company, retire the seed stock, and offer their own seeds. Monsanto went on a shopping spree.

As Bill Kte'pi explains in *Green Food: An A-to-Z Guide*, there are two kinds of expansion. One is horizontal, the consolidation of smaller companies that operate within the same sector in order to decrease competition. In vertical expansion, a corporation dominates all aspects of production. The corporate giants were practicing both types of integration; Monsanto moved vertically from chemicals to seeds, then horizontally through seed companies.

For a while it was dizzying. Under the header American Seeds Inc. (ASI), in November 2004 Monsanto acquired three Iowa seed companies—Crows, Wilson, and Midwest. In 2005, Monsanto announced that it intended to expand and spent $1 billion for Seminis, the world's largest producer of fruit and vegetable seed. In one fell swoop, Monsanto was the largest seed and biotech company in the world. It surpassed DuPont. On the heels of this purchase, Monsanto purchased NC+ Hybrids (supplier of corn and soybean seed, headquartered in Lincoln, Nebraska) and Emergent Genetics Inc. (cotton seed). Fifty-two million dollars later, in the fall of 2005, Monsanto owned Fontanelle Hybrids (Fontanelle, Nebraska), Stewart Seeds (Greensburg, Indiana), Trelay Seeds (Livingston, Wisconsin), Stone Seeds (Pleasant Plains, Illinois), and Specialty Hybrids. The purchases continued.

By 2006, ten companies controlled half the globe's commercial seed sales. Monsanto (United States), Dupont (United States), and Syngenta (Switzerland) led the vanguard. In 2007, the top ten seed companies accounted for 67 percent of the market. In 2009 Monsanto's market share for seed corn was 36 percent.

Monopoly is not a game.

A 2005 report detailed that the biotech industry as a whole lost $6.4 billion in one year. Since the mid-1970s it has lost more than $45 billion. So what's the reward for all this maneuvering? In many ways, the biotech industry in general (and Monsanto in particular) is like a giant start-up company. The corporate hope is that investors will, in the long run, see their efforts at conquest come to fruition. They hope to hit the jackpot sooner rather than later—and they may, as soon as enough of the world is conquered, enough farmers are disenfranchised, and enough of our food supply is lost forever.

And they may not.

That's what I'm voting on. That's what I'm praying for. That's the world I'm working toward—that every Frankenseed investor loses her ass and my neighbor farmers keep growing the seeds on which they've always fed themselves, their families, and their livestock.

4. BAD FOOD HAS BEEN FORCED DOWN OUR THROATS.
By now some of us know the dangers of toxic agriculture and the hazards of processed food, especially sugars, carbohydrates, fats, and high-fructose corn syrup. But another danger lurks.

The introduction of GM food has been a nightmare and folly. Without adequate or honest study, it has been approved by the FDA behind the smokescreen of one small but important phrase— "substantial equivalence"—which assumes that a novel food is as safe as the conventional food it replaces. The FDA website outlines the procedure for their approval of a bioengineered foodstuff. Developers submit a "summary of safety and nutritional assessment" 120 days before the GM food is marketed. The companies then compare a few key components, such as toxicity and allergenicity, against safe plants. "We monitor the levels of nutrients, proteins, and other components to see that the transgenic plants are substantially equivalent to traditional foods," said Monsanto's Eric Sachs. "If the levels are similar, then the GM experimental food is deemed identical for all practical purposes and no further testing is necessary."

Before the introduction of any new medicine or other powerful technology, long-term and thorough testing would seem to be in order: toxicology assessments, tissue cultures, multigenerational studies, allergy testing. Superficial testing does not account for the myriad possibilities of plant transformation and its potential to damage human health. In effect, the FDA's position is that there is no problem until a problem is identified.

Dr. David Schubert, a medical researcher at the Salk Institute in California, and William Freese of the Friends of the Earth published a paper in 2005 called *Safety Testing and Regulation of Genetically Engineered Foods*. "The picture that emerges from our study of US regulation of GM foods," said Schubert and Freese, "is a rubber-stamp 'approval process' designed to increase public confidence in, but not ensure the safety of, genetically engineered foods."

Nor have any labels been required by the US government—despite "right-to-know" campaigns by consumer groups and overwhelming support nationally for the labeling of transgenic products. Corporations prefer to avoid labeling since they understand that sales of GM foods would decrease if consumers knew what they were buying. By not requiring corporations to identify GM products, the FDA (more interested in the business of moneymaking than in the health of citizens) robs citizens of the right to know what's in their food. "Labeling is a situation where the FDA is officially charged with promoting biotechnology," said Jeffrey Smith, a visionary activist who has worked for years to bring attention to the menace of GM products and who is the author of the best-selling *Seeds of Deception*.

There is one way an eater can avoid GM food and that is to eat organic products. The National Organics Program, which writes the guidelines for inputs and systems that organic farmers may employ, prohibits the use of GM seeds. As Dave Hensen, director of the Arts and Ecology Center in Occidental, California, said, "Organics is one of the last lines of defense for all time." However, there is no way to avoid eating GM foods entirely because of contamination. *Acres USA*, a sustainable agriculture magazine, reports that genetic drift has resulted in the contamination of *all* US corn, for example.

5. Our food is hazardous to our health.

Actually we don't know this yet. The story of the perils of GM foods has not been told, in part because the mouths of the storytellers have been duct-taped shut and in part because the story is not yet known. The need for scientific studies has been ignored; those conducted have been suppressed.

The evidence of hazards in GM foods, however, is mounting. As Dr. Michael Hansen, senior scientist for Consumers Union, said in his 2002 lecture in Mexico entitled "Bt Crops: Inadequate Testing," "There is increasing evidence—from both epidemiological studies and lab studies—that the various Bt endotoxins (including those from maize, cotton, and potatoes) may have adverse effects on the immune system and/or may be human allergens."

Some of the adverse evidence comes from Australia, where a pea weevil threatens the field pea industry. Common beans, on the other hand, carry a gene capable of killing the pea weevil. When researchers in a ten-year project to develop a GM field pea tested this gene, it did not cause an allergic reaction in mice or people. However, when the gene was transferred to the pea, the new GM peas caused allergic lung damage in mice. What is significant about this study is that it indicates that a transformation happening during the transfer process may make GM food hazardous.

A variety of experiments have suggested caution is warranted on multiple fronts. For example, in spite of biotech's arguments that genetic transfer from crops to humans is unlikely, a study at Newcastle University in Britain found that DNA from GM crops could be transferred to bacteria in the human gut. Other experiments have intimated connections between GM foods and significant medical conditions. In rats fed GM corn and potatoes, scientists observed abnormal white and red blood cell counts, inflammation of the liver, and unexplained growths in stomachs and small intestines. In 1998, a scientist at the Rowett Institute in Aberdeen, Scotland, found that GM potatoes caused tumors and inflammation in the stomach lining of laboratory animals. The real world has even provided evidence of possible problems with GM foods. After Great Britain introduced GM soy, soy allergies rose 50 percent. Some Iowa farmers reported infertility (as much as 80 percent) in hogs fed GM corn.

In addition, the harmful effects of glyphosate (Roundup) are now rising in the American consciousness. Glyphosate kills weeds by shutting down their defense mechanisms, weakening them, and inviting infections by soil-borne pathogens; it is also linked to nutritional deficiencies in plants. It kills soil microbes, even the advantageous ones. Further, it does not break down quickly in the soil, taking anywhere from a few months to up to forty years. Clinical data implicates that glyphosate is, at "quite low levels," according to Don Huber, plant pathologist and professor emeritus at Purdue University, "very toxic to liver cells, kidney cells, testicular cells, and the endocrine hormone system." Additionally, the herbicide has been linked to miscarriages and premature births; it is implicated in Alzheimer's, Parkinson's disease, and even autism.

Once a friend said to us, "If you analyze food too much, you kill it." My outspoken husband had a quick reply: "Yes, if you analyze food too much, you kill it, but if you don't analyze it enough, it kills you."

6. OUR FOOD IS HARMING THE EARTH.

Not only is GM food probably unsound for the body, it leaves in its wake a host of other problems. The first is that insects and weeds evolve quickly to become superpests resistant to chemical controls. By 2006, eighty million acres were being planted with Roundup-Ready crops and being sprayed with Roundup. Then farmers began to notice that certain weeds were not killed by Roundup. Mare's tail—a tall weed with 200,000 seeds per plant—was the first I heard of. It became resistant to Roundup in only eight years.

During the past few years I've watched pigweed become resistant. At first only a weed or two would be left standing after a spraying, then the entire field would be dotted with pigweed. The solution, of course, is to spray even-more-potent herbicides, rocketing the farmers as well as the eaters ever further along a destructive path.

A 2009 study by The Organic Center concluded that "the most striking finding is that GE [genetically engineered] crops have been responsible for an increase of 383 million pounds of herbicide use in the US over the first 13 years of commercial use of GE crops."

Besides the increase of resistance to the herbicides, tolerance to herbicides may be transferred to weedy relatives of GM crops through cross-pollination. Canola is a brassica, a member of the mustard family. Many wild brassicas will cross with canola, even Roundup-Ready canola; when they do, the wild brassica, like the Roundup-Ready canola, will no longer be killed by Roundup. Again, more powerful herbicides will be used, turning most farms into greater and greater point-sources for pollution.

7. Our food annihilates pollinators.

"Where bees can live, so can man."
—Juliette de Biaracli Levy

The plight of our pollinators was outlined ingeniously by Stephen L. Buchmann and Gary Paul Nabhan in *The Forgotten Pollinators*. Farmer Frank Morton of Shoulder to Shoulder Farm in Philomath, Oregon, talks about the degree to which one attracts pollinators by seed saving. Morton's approach is to return as many processes as possible to the wild, looking to the garden as an ecosystem. Bolting and flowering plants, for instance, furnish continuous nectar, pollen, shelter, and prey for beneficial species.

8. Our food is nutritionally impotent.

The USDA identifies certain nutrients as vital: protein, carbohydrates, fat, fiber, vitamins (A, B-6, B-12, C, D, and E), as well as amino acids and minerals (thiamin, riboflavin, niacin, folic acid, calcium, iron, phosphorus, potassium, magnesium, zinc, copper). If every citizen followed the USDA's dietary guidelines, it promises, the United States would see a 20 percent decline in cancer, respiratory, and infectious diseases; 25 percent less heart and vascular diseases; and 50 percent less arthritis, infant death, and maternal death.

Corporate lackies will proclaim that a broccoli grown chemically (lifeless soil, drenched with cancer agents and endocrine disruptors) is essentially no different than one grown using organic practices (crop rotation, manures, legumes, compost, mineralization, microbialization). The studies beg to differ. Not only is there a difference, there's a big difference—up to a 100 percent difference.

A 2004 study of forty-three garden crops conducted by Dr. Donald Davis of the University of Texas–Austin found during the past fifty years reliable declines in six nutrients: protein, calcium, phosphorus, iron, riboflavin, and vitamin C. These drops ranged from 6 percent to as much as 38 percent (for riboflavin). Meanwhile, a review of USDA nutrient data on a dozen veggies by writer Alex Jack shows that since 1975 vitamin A has dropped 21.4 percent, calcium 26.5 percent, vitamin C 29.9 percent, and iron 36.5 percent. A study from Great Britain by Dr. Anne-Marie Mayer, looking at the concentrations of eight essential minerals in twenty fruits and vegetables, found consistent declines over the past fifty years—iron was down an average of 22 percent; calcium, 19 percent; and potassium, 14 percent.

Nutrients contained in a particular vegetable or fruit can be affected by many factors, including variety, maturity at harvest, and time from earth to table. A significant factor, however, turns out to be agricultural practices. New research is proving that food grown organically is more nutrient-dense.

High levels of nitrogen in chemical fertilizers stimulate quick growth and encourage a plant to take up more water. That results in higher yields, but less dry matter (the nonwater component of food) and consequently less nutrition and flavor per calorie consumed. Elevated levels of nitrogen reduce concentrations of vitamin C in vegetables like lettuce, endive, and kale, as well as in fruits like oranges and cantaloupe. Study results range from 6–100 percent increases in vitamin C in organic foods. Many inquiries demonstrate higher dry matter in organically grown crops, averaging 20 percent higher. In addition, analysis reveals higher mineral content in organic crops. Apparently compost delivers more nutrients than chemical fertilizers.

Bob Quinn, an organic wheat farmer from Montana, is growing kamut, the trademark (as well as ancient Egyptian) name of khorasan wheat, a variety supposedly found in the tombs of Egyptian pharaohs. His kamut boasts a higher nutritional profile than chemically grown wheat, including antioxidants, vitamins, and essential amino acids. It is mineral-rich, especially in selenium, zinc, and magnesium. "We've made these huge changes to make food cheap with no regard to nutrition," Quinn told *Acres USA*. "People say it's a great advantage having cheap food in this country, but if we take into account the cost of medicine from poor health, which is a direct result of our inferior and often toxic diet, it's not cheap at all."

With the spike in popularity and sales of organic produce, many chemical farmers donned tie-dye and peasant skirts, obtained organic certification, and began to grow what they call organic food, using organic inputs but without true belief in the underlying principles of organic agriculture. I know a chemical guy who grew his hair long. Wolves in sheep's clothing will be found out, however. As Eliot Coleman, farmer and author, said in a 2009 talk at Iowa's Heritage Farm, "the large growers may have changed their agronomy, but they didn't change their thinking, that their minds are still largely focused on how much they can produce rather than on how well it can nourish their customers." An organic monoculture is still a monoculture; its mission is profit. Coleman pronounces that "small is more beautiful" and praises the manifesto of small, organic farmers—"quality first, rather than quantity first."

In addition to organic agricultural practices, the role of *variety* in nutritional content cannot be denied. Breeding for size, color, and durability in shipping have contributed to the nutritional crash. "We're breeding for things other than nutrition," said Michael Pollan, citing a 30–40 percent decline in the nutrient value of American crops. "This is equivalent to us losing a whole serving of fruits and vegetables every day." In the past half-century, increases in yield have been paramount in the minds of breeders. As Davis said, "Emerging evidence suggests that when you select for yield, crops grow bigger and faster, but they don't necessarily have the ability to make or uptake nutrition at the same, faster rate." Lab tests of ancestral beans collected from indigenous people, for example, showed them to have eight to ten times as many antioxidants as grocery-bought beans.

9. OUR FOOD THREATENS DEMOCRACY.

Thomas Jefferson said he didn't think democracy was possible unless at least 20 percent of the population was self-supporting on small farms. These farmers would be independent enough to be able to tell an oppressive government to stuff it.

Instead, we are all increasingly helpless to provide food—not to mention good food—for ourselves. We are like infants needing to suckle at the bottle of corporations, which makes us dependent. And oppressed.

In principle, a democracy is one person, one vote, and every vote counts. Corporations want control, and under our capitalistic system, as Michael Moore pointed out in his documentary *Capitalism*, every dollar is a vote and the person with the most dollars wins. If corporations own our food supply, then they own us. The ability to feed ourselves ensures our freedom. As Eliot Coleman said, "Small farmers are a threat to the consolidation of absolute power."

a rind is a terrible
thing to waste

I APOLOGIZE IN ADVANCE for taking you on a short rabbit chase here, but to talk about seeds without talking about agriculture is not possible. The reclamation of seeds is the reclamation of food, and how we grow that food is important.

Plenty of seed savers use chemicals. But why? Why save ancestral varieties if you are not going to also save the ancestral farming practices that brought those varieties into being? The majority of America's vintage seeds were developed before the chemical era and grew fine without the use of synthetic inputs. So let's get one thing clear here, and that's terminology. You will never see me saying *conventional* agriculture when I mean *chemical* agriculture. Nothing about chemical usage is conventional, not if convention means what I think it means. The truly conventional, traditional, time-honored agriculture is the one that builds the soil.

One fall day in 2002, before he became a celebrity, I followed Joel "You Can Farm" Salatin around Craig and Debbie Hardin's Iraloke Farms in Salem, Florida. By then Salatin had published four how-to books on farming. My husband Raven and I had read a couple of them, and we found Salatin even more hilarious and entertaining in person than in print. "Organic farmers?" Salatin said. "People think the men wear ponytails and the women don't shave their legs and everybody runs naked through the woods on moonlit nights."

Maybe that's why he was looking perfectly conservative in his blue jeans and big white cowboy hat. There was nothing buttoned-down, however, about his doctrine. "The soil is the earth's stomach," he said. "In the soil, everything rots, rusts, and depreciates. The problem is that we treat soil like dirt and manure like waste."

"What we want to do is use nature as the template," he said. The farmer needs to mimic the symbiosis of nature. "We actually build forgiveness into the landscape with the checks and balances of biodiversity," he said. Chemical farmers don't do that. "The only tool in their box is technology. There can't be a natural cure—it has to be a high-tech cure."

He named the biological activity of *soil* as the key to tilth. "We're really in the earthworm enhancement business," he quipped and talked about one teaspoonful of healthy soil containing sixty million bacteria, three to five miles of fungal hyphae, a few protozoa and nematodes, and maybe a tiny insect or two. "That life supports healthy plants that support people's health," he said.

Industrial agriculture is responsible for the loss of two billion tons of topsoil a year. In the nineteenth century, just so you know, Iowa had fourteen to sixteen inches of topsoil. Today, it has just six to eight inches and more is being lost all the time.

It can take as long as five hundred to a thousand years to build an inch of topsoil. Granted, some farmers are proving that topsoil can be built much faster, as quickly as an inch of loam every five years, says Gardener's Supply Company founder Will Rapp. This can be done through high-carbon compost, broken down by microbes; through vermiculture; and through the coverage of bare soil—using mulch, cover crops, and plant-crowding.

Pre-Colombian natives in Brazil's notoriously infertile Amazonia created rich, black soils up to two meters thick that we call *terra preta*, or black earth, containing high levels of charcoal. Similarly, near Mexico City the Chinampa system created rectangular islands in shallow lakes, using lake sediment, mud, and aquatic vegetation. From these, farmers in canoes were able to draw four to six harvests a year. This birthed the Aztec empire and, according to anthropologists, proved itself the most productive form of agriculture ever. Then there are the organic farmers of China, Korea, and Japan who have reaped spectacular harvests from plots of land farmed for four thousand years, described in F. H. King's book, *Farmers of Forty Centuries*.

When I was twenty-two years old I went off to study organic agriculture with a man named Augustus Pembroke Thomson, although he went by A. P. A 74-year-old farmer in a straw hat and with kindness written all over his clean-shaven face, he grew apples at Golden Acres Orchard in Front Royal, Virginia. In June, just after I graduated from college, I read an interview with Mr. Thomson in *Mother Earth News* and wrote to him immediately: "My friend and I would like to pick apples for you." Mr. Thomson replied that it was not a bearing year, his expected crop was small, and he had just the space for two earth-loving stewards of the soil.

My friend, Irwin, and I arrived to a little trailer next to woods and beneath a huge windmill—eager to pick, eager to learn, eager to be surrounded by fruit. Early the first morning we were fitted with fifteen-foot ladders and pick-sacks, buckets which hung from the neck and strapped across the back, which held about a half bushel and filled quickly. First you pick apples you can reach from the ground, Mr. Thomson told us, then position your ladder and pick as you climb. The ladder needs to rest on a V-shaped limb for support. When the sack is full, you climb down the ladder, unhook the tote's canvas bottom, and gently roll the apples into a big wooden bin. Up and down, around and around, until the tree is done. "Don't worry about anything on the ground," said Mr. Thomson. "We'll pick up the windfall later to make apple cider vinegar."

"Our first day picking is wearily over," I wrote on September 23, 1984. "Our backs and shoulders are sore, face sunburned, arms scratched. We picked long-stemmed Golden Delicious, the hardest to come off and the easiest to bruise. We filled four bins—that's eighty bushels, about minimum wage."

Within a couple of days I was taught to drive a tractor. "Why are women led to believe it's beyond their capabilities?" I wrote. Mr. Thomson appeared at the trailer early one morning and loaned me three books to read. "Then we'll talk farming," he said.

Forty years before, Golden Acres was a deeply eroded and unproductive family farm. Mr. Thomson was in the Navy, in Pearl Harbor, and he happened upon the farming classic *Pleasant Valley* by Louis Bromfield. The book made sense and Mr. Thomson began to read everything he could get his hands on about organic agriculture. He began corresponding with Sir Albert Howard, an English botanist considered the father of organic gardening and author of *The Soil and alth*; and J. I. Rodale, founder of Rodale Press, publisher of *Organic __rming and Gardening* and *Prevention* magazines, and popularizer of the word *organic*. In an interview for the *Washington Post Magazine* in 1985, Mr. Thomson said, "You must understand: it was a rare thing in those days. I was fortunate to meet up with the real pioneers."

Mr. Thomson returned to his family home full of new knowledge about organic agriculture and began to nurse its land back to health. He spread chicken manure and planted green manure crops, bromegrass and sweet clover; he tilled them in, sweetening the soil. In the third year he planted apple trees. When I was there he was still doing regular soil tests in a tiny, cluttered laboratory and adding colloidal phosphate, calcium, and other nutrients yearly. He sprayed Norwegian seaweed and fish emulsions during the growing season. He spread cattle manure. He hand-raised Red Wiggler earthworms.

Irwin and I looped through Red and Golden Delicious, Winesap, and heavy-hanging York, one to four bins daily. We befriended the few other workers, including Mrs. Thomson, who put in long hours grading and packing. One of the pickers was a guy from Washington State named Bob, a wanderer who smoked the thinnest joints I've ever seen. He was twenty-seven. He said that nothing mattered much because the earth was not reality anyway.

Even when I was dog-tired, the thirty-five-acre orchard never lost its majesty. The grass was full of blooming vetch, alfalfa, red clover, and Queen Anne's lace. The trees, in orderly rows, were spangled with colorful apples like ornaments. We paused to eat countless apples without peeling or washing them, shooing away bees. "It is an indescribable pleasure to walk out and pick breakfast," I wrote.

I was a sorry picker. Mostly I wanted to climb an apple tree with a book and read all day. I wanted to gather pokeberries down by the track and squeeze them into ink. But that month in the Shenandoah, I received a crash course in how to grow food right.

Mr. Thomson didn't call his style of farming organic. "For a long time," he wrote in a declaration of principles that he shared with me, "I have felt that the terms 'organic' and 'natural' were being abused so we adopted a term we could define and defend if need be; that being 'biologically grown.'" Chemicals treat symptoms of poorly managed soil, but the most productive, healthiest way to grow food is to love the soil. There are as many living organisms in a single cubic inch of healthy soil as there are human beings on the planet, he said. It's an integrated, pulsating, living, breathing system.

Thomson was a philosopher as much as anything. He believed in mimicking nature. "When you sit by a flower, don't be as a person, be as a flower," he said prophetically. "When you sit by a tree, don't be a person, be a tree; and when you do this millions of signs are given you. It is a communion, not a communication; nature speaks in thousands of tongues but not in language."

He was fanatical about remineralizing degraded soil with minerals like calcium and magnesium, and trace minerals like boron, chromium, iron, and zinc. He believed that this mimicked the action of glaciers eroding mountains and building soil and that the minerals fed microorganisms in a biological and restorative matrix, multiplying the health-giving aspects of his crop.

"All life in its broadest concept exists in a state of dynamic tension of the opposites, an unbelievable exquisite balance and harmony. There must be balance in the cell, the soil, and throughout the universe." To this end, Mr. Thomson was also channeling what he called cosmic energy into Golden Acres. In the orchard, he had erected two metaphysical energy receptors—metal towers holding aloft small chambers of copper tubing filled with minerals gathered from around the world—based on ancient Irish towers and the research of Dr. Philip Callahan, then of the University of Florida. A small sign read ORTHOMOLECULAR MULTIWAVE OSCILLATOR. The towers elevated a pair of antennae that worked like the old crystal radio receivers. The minerals vibrated on the same frequency as the apple trees, bringing added energy. Christian monks in Ireland between the fifth and seventh centuries built ancient stone silos that contained metallic ore that could collect and resonate the sun's radio energy, believed to increase the harvest.

Kooky as they sounded, Mr. Thomson's odd theories seemed proven in every perfectly shaped, officially nutritious, and absolutely delicious apple that came in from the sun.

At night I read the farming literature supplied by Mr. Thomson. I listened to audiotapes of Dr. Carey Reams's *Theory of Biological Ionization* and Philip Callahan's work with insects and paramagnetism. I took notes and made plans.

I was abhorred by all I learned about chemicals. One day Mr. Thomson's son told me that I should see the groundhogs at the next place, which was not organic. "I've seen them with no hair on their bodies, stunted legs," he said. "It's awful." The younger Thomson echoed his father in blaming the deformities on the pesticide endrin, which causes leukemia and was used to eradicate mice in orchards. He told me that apples in chemical orchards were sprayed many times with malathion and paraquat. The ground was heavily herbicided and artificially fertilized. For commercial apple juice, the apples were cursorily rinsed, then additives, like aluminum sulfate to dissolve the unsightly pectin, were used in bottling. "The end product is not apple juice but a gallon of poison with a mild apple flavor," he said.

One day Mr. Thomson told me about a chemical peddler who wanted to know why Thomson bothered with organic farming. Mr. Thomson replied, "I want the comfort of knowing that if a little child picks up one of my apples without washing it first, he won't be in danger of being poisoned." He cursed apples "grown with chemicals carrying the skull and crossbones, immersed in coal-tar-derived skin preservatives, sprayed with fungicide-impregnated, petroleum-derived wax and probably containing systemic poisons administered to the tree roots."

Mr. Thomson also shared his vegetable garden with us—carrots as big as wine bottles, tomatoes, butternut squash. Once, coming out with a handful of huge beets, we met Mr. Thomson going in. "Phosphate and aragonite," he smiled. "That's what does it."

"I just want to teach people what I know," Mr. Thomson said, "and inspire them to try new ideas themselves." What I learned from Mr. Thomson is that the Agribusiness Total Chemical Age, as he called it, begets pollution, soil degradation, health hazards, loss of biodiversity, erosion of nutrition, and pauperization of rural communities. It took me many years, and a lot of twists and turns, but I'm finally building some soil.

You must understand: I was fortunate to meet up with one of the real pioneers.

losing the conch cowpea

Soon after I moved to Sycamore, I began to package seeds in tiny manila envelopes labeled Hoedown Organic Farm and ship them here and there, a dollar a pack. By the time I was twenty-four, in 1986, I was exhibiting heirlooms and novelty plant seeds at the Florida Folk Festival, an extravaganza held every Memorial Day weekend on the banks of the Suwannee River, at Stephen Foster State Park in White Springs.

Old-timers would hover lovingly over my table, peering into baby food jars of wild coral bean, the only truly red seed I know; droplets of tobacco seeds; knots of brown cotton. They would lightly shake vials of Velvet bean, four o'clocks, Red okra, Moon and Stars watermelon, Soldier bean. I remember one slight Cracker gentleman in a cuffed shirt and a ball cap. "Bless my soul," he said. "This is Conch pea."

"You know it?" I asked.

"Oh, hell, I've planted a million of these. We grew 'em when I was a boy. It lies close to the ground when it grows, like a sweet potato vine. Haven't seen them in years."

"I never heard of it until recently," I said. "I saw seed advertised in the *Florida Market Bulletin* and I bought some and grew it. I can't find it in any of the catalogs.

"It's a real old breed," the gentleman said, adjusting the waist of his pants. "It used to be very popular."

"Did you hear where it came from?"

"No."

"It was supposedly found on a Florida beach with debris from a shipwreck."

"I'll be damned," he said. He bought a couple of packs.

Before long an older woman was slipping her reading glasses on her nose. "We used to grow this Cowhorn okra when I was a girl," she said. "It's the one with the long pods, right?"

"That's right."

"What happened to it?" she asked.

"I don't know. It got discontinued for some reason. It's a wonderful okra as far as I'm concerned."

Or these Jacob's Cattle beans. Or German Pink tomatoes. "I'm so glad to see you again," said a heavy man, a retired farmer with pink cheeks and wire-rimmed glasses. "I only came to the festival this year because I thought you might be here." He bent to the seeds like a scientist, working his way through the envelopes and jars, as if they restored something unnamable to him or as if to handle them was a right that he had been denied most of his life.

If that Sycamore existence had been my mythology and if it had been feasible, I might still be in north Florida in an ever-more-microbial, leafy, and rich garden, experimenting and developing. But I was called away. One day while I was mulching our Southern apple trees next to the road, a man stopped for directions. He was a dozen years my senior, handsome and interested in homesteading. We began to date and then live together. I was married and had given birth to a baby boy by the summer of 1988.

In a fading photo I have of that year's folk festival, I am sitting behind a table lined with seeds, chatting with a customer, cradling our newborn son in the crook formed by crossing an ankle over a knee. Silas would have been two weeks old.

After that, there was sadness. My fragile marriage didn't last the year, and for many years to follow I had no garden—no time, no land, no energy. I drifted far from the enchanted young woman among flowers who kept bees and saved seeds.

I was, however, given a second chance. It came after Silas went away to college. When your son leaves, someone told me (preparing me for loss), fill the space with love. I filled the space of Silas's absence with old loves, especially seed saving, and I took up where I had left off eighteen years before.

Right away I thought about Conch cowpea. Over the years, moving from here to there, a single mom for most of the time, I'd lost all trace of my supply. Now I went looking in seed catalogs. I looked online. I looked in the *Georgia Market Bulletin*. I renewed my membership with Seed Savers Exchange and looked there.

Conch cowpea was nowhere to be found. In the time I had turned my head to rear my son, it had disappeared. Was it gone for good?

Cowpeas are important drought-tolerant legumes adapted to the tropics. Some cowpeas are bushes; others put out runners (sprawling vines, although not as long as those of pole beans); others are more demure half-runners. In a garden bed, cowpeas will spread four feet in each direction. In a subtropical climate where northern beans are not wont to grow, cowpeas are a vital source of protein. They produce well in sandy, poor soils; enrich the soil by fixing nitrogen; and with luck supply a family with protein all winter. They can be eaten as snaps when young, as fresh peas when mature, or as hoppin' john when dried for winter use. In addition, the dry vines are cow fodder. Traditionally they were grown in conjunction with field corn, which is why many varieties are called Corn-field beans. Cowpeas are divided into two main categories, explained to me by my farmer friend Vickie Carter—crowders and creams. (These are further separated into Purple Hulls, Long Pods, and Forage peas.)

Crowders are named for their habit of crowding each other in the pod, to the point that the peas are squared-off. Crowder peas usually exhibit some coloring and may be brown, black, or speckled. The color often concentrates around the hilum, or eye; though usually black, the peas may also be pink, brown, or tan. When cooked, the pot liquor of crowders is dark. The bloom is usually violet.

Cream peas, on the other hand, mostly produce white or partly white seeds and a light pot liquor. Some varieties have a spot of color around the hilum, or seed scar; black-eyes are cream peas. The bloom is white.

Butter beans, another protein-rich Southern staple, are not cowpeas. They are what can be bought dry in the store as lima beans, although Southerners traditionally ate them fresh more often than dried. Butter beans also grow either on bushes or on vines, in the garden or in the cornfield.

During one of my peregrinations someone handed me a manila envelope. It contained two agriculture newsletters from 1958, and both of them were the *Georgia Market Bulletin*. One was from February and the other from June. The newsletters were yellowed and crumbly and I went looking through them, wondering about farm life in 1958—when the chemical era was firmly in place but more folks still lived agrarian lives and local cultures were more intact, if not more functional. Seed for sale was advertised under the generic column "Seed and Grain" with the exception of "Beans and Peas." That's because beans and peas have always been crucial to the household economies and foodways of the southern United States.

In the advertisements from 1958, the unbelievable diversity of the varieties surprised me. The peas were sometimes offered in large quantities, such as 350 bushels, and sometimes in cups. (Unable to distinguish which words are descriptors and which are names, I've capitalized everything in this list.)

Old-fashioned Large White Half-runner bean
Black & Brown Cornfield bean
Speckled Cut Short Cornfield bean
Brown-striped Cornfield bean
White Tender Creaseback Cornfield bean
Black Crowder pea
Brown with Dark Purplehull Crowder
Red-spotted Crowder
White Crowder
White Brown-eye Crowder pea
Red Speckled Cream & White Crowder pea
Cream Crowder
Old-time White Pole Butter bean
White Creaseback Half-runner
Purple Blossom Brown-Striped Half-runner bean
Old-time Speckled Half-runner bean
White Half-runner Garden bean
Brown-striped bean
Little Pink Peanut bean
Tender Hull White
Purplehull
Little White Lady pea
White Bunch Butter bean
Henderson White Bunch Butter bean
White Garden Bunch bean

Blue Java pea
Purplehull Pole bean
Blue Pole bean
White Black-eye
White Acre pea
White Acre Conch pea
Brabham pea

Look at how many there are, a variety for every microclimate. A variety for every family. A variety for every use. Is Henderson White Bunch the same as White Bunch? Is it different because it came via a family of Hendersons? Was it grown in a place called Henderson? If we did genetic testing, would all of these be different?

Scrutinizing each name, it is obvious that each pea has a story—of where it came from and how it was grown and used—and we will never know these stories.

Notice in the list that a conch appears. It's called White Acre Conch. Is this a mix-up of the names of two popular peas? Is it a cross? Is it something new that reminded the farmer of both White Acres and Conchs?

I'm also especially fascinated by the Blue Java pea. Not long ago I called a man, Walker Ogden, to see if the corn a Mr. Ogden had given a Mr. Gore was his family's heirloom. Walker did not know. The family no longer had a vintage corn, but it did have a pea his family had been growing. They had always called it the Javie pea.

"How do you spell that?" I asked.

"Honey, how would you spell *Javie pea*?" Walker called to his wife.

"J-A-V-I-E," she said.

Although he had not grown it in a few years because he was working off the farm, he had plenty in the freezer and planned to grow it again. He would gladly share a start.

We've only begun to leaf through the cowpea lexicon. It goes on and on: Calico Crowder, Green Acre, Dixie Lee, Stowwood, Trinkle's Holstein, Smallpox, Ram's Horn, Polecat, Cuckold's Increase, Browneyed Sugar, Buckshot. Add to this list the varieties described in Charles Piper's 1912 tome *Agricultural Varieties of the Cowpea and Immediately Related Species*. Add to it the more than one hundred cowpeas in the *Garden Seed Inventory*, a book compiled by Kent Whealy of the Seed Savers Exchange that collates information about nonhybrid seed availability from commercial sources in North America. We could fill pages and pages with the names of cowpeas alone.

I made a call to our local feed and seed. It still offers fifteen varieties of cowpeas, including Mississippi Purple Hull, Texas Cream 40, White Acre, and Dixie Lee, which shows how important field peas are to Georgia growers. But hundreds and hundreds of varieties are no longer offered commercially. The clerk told me that many of the old-timers ask for Red Ripper, a variety I obtained from a local gardener, Harry Mosely, who obtained it from Short Reeves. "It's a real good pea," Mr. Mosely told me. "It'll turn your juice real black."

We've lost a lot of cowpea varieties in modern history, and cowpeas are only one foodstuff. Expand this diversity to every garden crop grown in this sweet earth. We are besieged by loss.

– 8 –

hooking up

SEX IS WHAT'S HAPPENING at Will Bonsall's farm, shameless sex all over the place, even if it's rutabaga sex.

Bonsall is a famous seed saver who manages a seed-saving endeavor that he calls the Scatterseed Project. He was getting so many visitors to his farm, where the seed saving takes place, that he decided to schedule an open house once a year in order to preserve his privacy the rest of the year. Raven and I happen to be in Maine for my work—I'd just finished teaching a writing course on the climate crisis—and we have arrived at Bonsall's farm for the tour. The only advertisement is on the butt-end of a cardboard box on a closed gate with KHADIGHAR FARM/2 P.M. lettered on it. Angles of a wooden house are visible through shaggy trees.

We park on the road. A woman in the lane is busily moving vans, one of which has a single bumper sticker: BIOTECHNOLOGY: GIVING POLLUTION A LIFE OF ITS OWN. She calls down a hello, says they'll be ready at two, she has to take somebody to fiddle camp. It's a quarter to two. We get out and stand in the damp road. The trees—maples, hemlock, fir—are dripping.

Raven brings lobster from the car, left over from the workshop's final banquet. He eats the last of it and rinses his hands beneath our water bottle. "Look left," I whisper. Six turkey poults, days old, cross the road and vanish into the woods, where the fern-covered ground is like a rainforest.

At two on the dot Bonsall descends to the road barefoot, wearing large blue-denim pants with pockets on the sides, held up by black suspenders, beneath a homemade light-denim tunic. He has a long gray beard and a widow's peak. He asks if we're all the people there are, and we say so far. Just then another car arrives and two more people get out.

"I think rain is keeping the others away," Bonsall says as he herds us up the hill. Clover and plantain grow among the grasses. Between the road and the house is the first garden.

"I see we have the same kind of tractor," Raven jokes. It's a shovel, left out in the weather. Bonsall doesn't laugh. Maybe he doesn't hear Raven.

"Our primary focus here is twofold," Bonsall begins. "Number one, I've always been interested in self-sufficiency. And number two, doing it within a vegan framework."

I don't understand. "Vegan?"

"Organically, without using animal products. Using green manures. We have one exception, composted cow manure. I call our system deep conventional."

Bonsall's desire to be self-sufficient prompted him to call his farm Khadighar, formed from two Hindu words—*khādī* meaning "a coarse, hand-woven cloth" but coming to mean "a movement for self-rule," specifically the Indian independence movement led by Gandhi; and *ghar*, meaning "house" or "homestead."

"To me it means 'homespun,'" Bonsall says. He stops and grins. "People think we're Hare Krishnas. On some kind of a trip."

Bonsall's eyes are very blue, set in bronze skin. "In the beginning, I wasn't even aware of heirlooms. I was simply interested in crops conducive to self-sufficiency. Naked-seeded pumpkins. Fiber flaxes. Long-string paste tomatoes."

One evening during the 1970s Bonsall attended a gardening talk and afterward someone handed him a Seed Savers Exchange catalog. The organization was in its second year and Bonsall joined immediately. Now he maintains the largest collection by far of any member, about 1,300 varieties. Where, I am wondering. The garden I am looking at is a nice garden, but it isn't the outpouring of verdure and life that I usually see in gardens. The plants are on the thinnish side.

When he first started curating seed, Bonsall's focus was Maine heirlooms, geographical within five hundred miles, and he spent a lot of time collecting in the Northeast. One of the varieties he found was Byron Yellow Flint corn.

This is as good a time as any to explain that there are a number of kinds of corn. Flints have outer layers that are hard as rock and don't bear indentations. Indian corns are flints, and popcorn is a variant. With more soft starch, dent corns have distinctive depressions along the sides of the kernels. Shoe peg corn means the kernels occur in zigzags, as opposed to rows; Country Gentleman is an example. Additionally, some corns are sweet corns, bred for eating fresh; some are field corns, usually dried and used for animal feed.

As the story goes, Bonsall heard that a Mr. Mosher of Wilton, Maine, had an old corn and went to visit him. The man was practically on his deathbed. He said he had no more of that corn, hadn't planted it in years, but after a while he remembered something. He sent the woman he called his housekeeper to retrieve a shoebox under his bed. Bonsall still remembers the housekeeper's name—Olive. She returned with the shoebox that contained one single ear of corn fifteen years old.

A seed has a lifespan. Longevity, the ability of a seed to survive in a dormant state, depends on conditions—moisture, temperature, kind of seed coat—as well as other, more mysterious and little understood, factors. Sometimes we don't want a seed to be long-lived. Weed scientists diligently study the power of weed seeds to persist in the soil, for example—they talk about half-lives, the half-life of common pigweed and lamb's-quarter seed in the soil being just over one year. Mostly we want crop, not weed, seeds to persevere.

Many claims of long-lived seeds are false, but viable seeds six centuries old of a species of canna were found in a tomb in Argentina. Canna seeds had been inserted into green walnuts, which became rattles once the nuts matured and dried. The walnut shells were carbon-dated as 600 years old.

In general, however, for the purpose of home gardens, once a packet of seed is opened, seed will keep only a matter of years—cucumber seed for seven, tomato three, salsify two. The conservative estimation of longevity for household storage of corn is a only a few years. Knowing this, Bonsall brought the old man's corn home and was extra careful in planting it. Unbelievably, some of the seed germinated!

The corn is a very early maturing flint variety, long ears with cobs like knitting needles and amber kernels that grow in eight rows, becoming twelve rows at the top. Bonsall called it Byron Yellow Flint corn and learned that it was probably a corn originally developed by the Abenaki Indians of the Northeast.

"I've sent out dozens and dozens of packets," Bonsall says.

Our little group moseys on up to a second garden, which is in better shape. It is mulched with shredded leaves. "We have a fetish about naked soil," Bonsall says. "To have bare soil is like staring at a woman in labor. Put a blanket on her and respect her." A pitchfork has been forgotten at garden's edge, and an axe lies rusting on the ground.

Opposite the barn, a verdant, emerald vine grows on a large trellis— kiwi, sassy with unripe fruit. Bonsall promises they will be delicious. "People of opposite genders should not be allowed to eat these in the same room," he says.

At the second garden sex education begins in earnest. "Sex is a good thing," Bonsall says. Hundreds of thousands of new varieties come into existence by sex. But what if we're trying to keep the same variety? Then we have to control who has sex with whom.

"You don't get pregnant sitting next to somebody on the bus," he continues. "But plants do. So we have to be careful which plants are in bed together. Sometimes we have to make sure a plant has sex with itself. Think of it like this: variety is just a euphemism for race. What we're doing here would be blasphemous in humans. We're Nazis here. We're doing vegetable racism. This serves our purpose, not theirs. We're not after yield. We're after purity."

Standing in front of rows of vegetables, no two rows alike, he talks about boy parts and girl parts. "If I say stamen and pistil and anther, we're off on another planet," he says. Soon, however, he's talking heterozygous and homozygous, until a bunch of Bs are buzzing around: Big B, little b. Big B, Big B. Little b, little b. It sounds like some kind of language poem. It sounds as if I'm on Jupiter. I know right then I need to borrow a genetics text.

I ask him to explain F2 and F3 and so forth. I understand F1, first-filial cross, meaning the hybrid sons and daughters, so to speak, of two parents who happen to be different.

"F2 are grandchildren," he says. "You cross two of the F1s to get them. But listen, F1s are what you have when a plant breeder doesn't finish his work."

"Meaning?"

"If F1 was looked at as the beginning of a breeding program, that would be different. The breeder could take seven more years and stabilize it. But breeders want a certain variety on the seed rack next year. When they start finishing their work, then the market will have some beautiful open-pollinated varieties." He pauses. "I will not grow an F1 hybrid. I will not have one on the farm."

This is as good a place as any to stop for a rest and explain to you why seven years is not a magic number pulled from a hat. Tom Stearns, an entrepreneur and seedsperson who started High Mowing Organic Seeds in Vermont, which is dedicated to selling only organic seed, explained it to me. He and I were on a seed-saving panel once at a Georgia Organics conference. Afterward, when I had a question that stumped me, I felt as if I could call him up and get a straight answer from an expert, a person infinitely more knowledgeable than I.

"Well, it's not seven years," he said. "It's seven *generations*. If you're working in a warmer climate such as where you live or in a greenhouse, you may get two generations in a year."

"Okay, seven generations," I say. "But why?"

"When you make a cross," he said, "and save the seed, it takes a number of years for the seed to grow true. Each year your seed gets closer and closer to the characteristics you desire. Basically it's a matter of statistics. On average, the process takes about seven years."

I was still in the dark.

"Let's say that with the first generation you're 20 percent of the way toward the characteristics you desire. The next generation, you're 30 percent, then 40 percent, then 50 percent. In seven generations, you'll be 95 percent of the way toward true."

"Is it always seven generations?"

"No," Tom said. "If you're crossing two similar varieties, let's say two red slicing tomatoes, both determinate, you may stabilize the cross in three or four generations. But if your cross is wide, if you're crossing a red slicing tomato with an orange cherry, for example, it will take seven generations and maybe more."

"Now I understand," I said and thanked him.

There are different places to draw the line about what kinds of seeds you will embrace and those you will not. Bonsall draws the line at hybrids. Even as he avows this, somewhere at Khadighar some plant is hybridizing, without a doubt. I'm sure it's happening, accidentally. It happens all the time. Folk growers have been producing happenchance hybrids for centuries, hence all the agricultural diversity to start with. No doubt about it, thoroughbred is inbred. But for Bonsall, inbreeding maintains purity.

Many small and organic operations draw the line at GM and they don't think twice about planting hybrids. Johnny's Selected Seeds, to name one example, which caters to the market gardener, appears to offer more hybrids than standard varieties. One afternoon while in residence at Pace University in 2010 I was touring Stone Barns, the Pocantico Hills, New York, estate of David Rockefeller that has become an apprentice farm and home of the most famous farm-to-table restaurant in America, Blue Hill at Stone Barns. Executive director Jill Isenbarger told me that the farm is a center of experimentation for new hybrid seeds; in fact, she could only think of two heirlooms being grown that season, Otto File corn and Panther soy. To grow hybrids is to accept that most seeds are a product of hybridization—"to see what happens," says the Stone Barns newsletter, in the "spirit of artful innovation."

But for Bonsall, hybridization equals industrialization, and he isn't willing to go there.

Bonsall is a regional curator for the Seed Savers Exchange. He has chosen to keep alive some very difficult seeds. "Beans and tomatoes—that's the seed most people save. For one thing, they're sexy crops, all annuals, all self-pollinators," Bonsall says. "But I specialize in two-hoop crops."

He doesn't wait for one of us to ask what he means.

"You have to jump through two hoops to get the seeds," he explains. "These plants are mostly biennials. The first year you plant the seeds, then you must overwinter the plant. They go back into the ground, flower, and make seed."

To trespass against the rules of biennial reproduction would be to ruin a lineage. And rules there are aplenty, enough to boggle a mind. Here are some of the rules. Plants cross-pollinate. Even though they look and taste so different, beets and chards cross, since both belong to the same genus, species, *and* subspecies—*Beta vulgaris vulgaris*. Beet and chard pollen is fine, evolved for wind pollination, and these plants must be separated by two to five miles for absolute purity. Alternately, their flowers may be bagged.

"Bag?" I interrupt.

"To cover a hand-pollinated flower with a thin bag in order to prevent cross-pollination. If a plant is insect-pollinated, it's more difficult to maintain seed purity."

"Got it," I say.

Bonsall continues his litany. Roots stored overwinter in cellars must not freeze. They must be replanted the next spring back into the ground before they rot. "All these rules are kind of difficult to follow," Bonsall says. The difficulty explains why fewer people save the seeds of biennials. Bonsall curates Brussels sprouts and leeks for the Seed Savers Exchange. "For better or for worse," he says. "And mostly worse." He is also keeping alive a variety of Swiss chard that's from a former East German pre–World War II town. "The people who grew it are rotting in some mass grave," Bonsall says. He pauses, not long enough for anyone to ask a question.

"Gulp." It is not a real gulp. "I better not frig up. There's a very great danger in so much genetic base in one place. If this house burned down, there would be a hell of a lot of extinction."

"So many of the varieties are doubtlessly identical," I say to Bonsall. "They simply have different names. Why not test them and reduce the collection?"

"The DNA testing is costly and inefficient. Once some testing was done on the Seed Savers Exchange's potato varieties. The idea was to throw the repetition away and keep only unlike varieties. The best they could ever find out was, 'Yes, this potato really is different.' They could never find a test that said, 'Yes, it really is the same.' Essentially, then, a seed saver can never throw anything away. What the potato people were looking for was simplification," Bonsall says. "The virtue in simplicity is that it's easier. It's also very dangerous. There is safety in complexity. There is always strength in complexity. I am wacko into diversity," Bonsall says. "The more, the merrier."

Bonsall changes course. "Speaking of simplicity," he says. "People used to say that I'm living a simple life. But the supposed simple life is very complex. Every single hour I have to stop and rethink, 'What am I doing?' The options are constantly changing. This is a very complex life."

We finally head uphill to the main seed gardens. Past a fencerow a splendid wooden wheelbarrow lies forlornly out in the weather, sun and rain and snow. Bonsall said he built it with an apprentice. They went to a museum for the design and used ash for the spokes, elm for the wheel, and cedar for the box.

The garden is in complete disarray. A week earlier Bonsall hired a young man to come and try to make sense of a few things, he says.

Rows are weeded about twenty feet in, enough for viewing, and the rest is left to wilderness.

The potato collection is over seven hundred varieties. We can't really see the potato vines. They are swallowed by weeds, which Bonsall admits got out of hand with all the rainy weather that summer. "They'll be okay," he says. "We just want them to survive and make a few tubers."

"I'm amazed at how few of the varieties you can get commercially anymore," he continues. "There are a few where this is the last place on the planet you can get them. And frankly, that scares the shit out of me. If I lose them, they will be gone from the planet."

Frankly, that scares the shit out of me too, because I can't even see the potatoes. I spot some sickly, frail plants dog-deep in brush.

Bonsall admits that he's overwhelmed. He's fifty-nine now, unable to work as hard as he once did. Income from seed-sample orders and occasional grants are not enough to maintain all this diversity on a sustainable basis.

"You maintain every single variety that comes your way?" I ask Bonsall.

"This is Noah's ark," he says. "I'm Noah. I'm not God. I don't get to make the decisions about what to put on the ark. Maybe later we'll say, 'Whoa, it's a good thing we didn't lose that during the Monsanto era.' A variety may not be hardy here but it might be great someplace else. It may have a very valuable gene. I don't want to put values on these things. It's not up to me to make that judgment."

Bonsall's a man of metaphor, and he finishes his hour-long monologue with a flourish. "I'm trying to juggle a couple thousand balls here. I can't let any fall on the ground, even though I may not be great at juggling. I'm fighting a battle I'm doomed to lose," he says. Then he shrugs lightly. "But that's what life is." He hopes seed savers "develop their skills with the more challenging crops and join me in protecting these undercurated crops." As for the potatoes, in 2008 the Seed Savers Exchange began a project to back up the potato varieties at Khadighar with laboratory tissue cultures for long-term storage.

We leave Bonsall juggling over seven hundred varieties of potatoes, over fifty of Jerusalem artichokes, over four hundred peas, over fifty radishes, and five leeks—all up in the air in his big orgy of a garden.

– 9 –

sylvia's garden

AT THE ENTRANCE to Sylvia Davatz's garden grows the largest known white oak in Vermont. In the oak's branches robins are singing and all around, the green of summer is its own animal.

Before us, in an area a few hundred feet square, this elegant woman maintains about 150 varieties of garden plants. A series of paths thread between narrow raised beds configured to maximize sunlight and space. Sylvia wins the "neatest garden" award. Her beds are planted meticulously and evenly. Twelve lettuces spangle a weedless section, next to a chopping block of onions, next to a chessboard of cress. Although intensely organized, the garden still manages to go nuts, and a strange, dazzling red-leaved plant has sprung up in crannies and edges. "Red orach, similar to lamb's-quarter," Sylvia says. (Both are in the *Chenopodium* family, cousins to quinoa.) She pulls off a crinkled, arrowhead-shaped leaf and hands it to me. The leaf tastes like spinach, with less oxalic acid.

Sylvia was born in Switzerland. Her father was an American representative for a company that made surveying and other precision instruments. Sylvia grew up in the United States but her family returned to Switzerland when she was sixteen. "When I was a kid I read *Swiss Family Robinson*," she said, "and I was absolutely smitten by the ideas of self-reliance and resourcefulness." After eight years in Switzerland, Sylvia flew back to the United States and tended her first garden in 1978 in Milton, Massachusetts. As the years passed, she began to see pet varieties disappearing from seed catalogs. Seeds were being lost. They were being displaced by hybrids, taken over by corporations. To save them was a basic but necessary skill, and Sylvia began to see herself as a steward of seeds.

"A lot of people are just beginning to garden and already thinking about seed saving," Sylvia continues. "This gives me a lot of hope, since it means awareness is growing about how important the work is. I came to it much more gradually."

"Was it difficult to switch from gardening to seed saving?" I ask.

"There's definitely a shift in awareness of timing and the need for observation. It feels like learning a new Romance language: I already know French but am learning Spanish. The grammar is the same but the words are different."

In this moment I'm distracted, trying to listen to a very thoughtful gardener and, at the same time, experience every inch of her garden. We come to a bed of True Red Cranberry, a pole bean she obtained from Abundant Life Seed Company in 1997, when the variety was rare, before it made a grand comeback. "I tell new gardeners to start with simple things, self-pollinators like peas and beans. All you need to do to save legumes is let them dry on the vine."

We stop at a drift of zucchini. "I grow only one variety at a time," says Sylvia.

"Because they cross easily?"

"Yes, it's a challenge to keep varieties pure. And I plant only one species of winter squash a year."

"Which one, may I ask?"

"Red Kuri." Red Kuri is a thick-skinned, deep-orange winter squash that is popular in Japan. "Interestingly, without my knowing it, my sister in Switzerland's favorite squash is Pôtimarron, which looks identical to Red Kuri. This year I decided to grow both, to see which is best. But only one is here. The other is in a friend's garden."

It is soon obvious that Sylvia is an experimenter, an emblem of a fine mind.

"My latest project," she says, "is the search for two outstanding onions, a yellow one and a red one, that will be particularly suited to our area." She explains that onions are biennials, meaning they flower and produce seeds in their second year. So to get seeds, the root must be stored for the winter and then replanted, after which the plant will flower, then set seeds. "Last year I grew fifteen different onions. I stored them in the root cellar. I have records of when they broke dormancy."

"I don't see fifteen different onions here." I look around.

"Oh," she smiles. "I parceled them out to friends to grow. I'm growing to seed in different gardens."

"Are these friends all over the country?"

"Most are right here in Hartland, Vermont. I want to be profoundly local." I like that: *profoundly local.*

We pick our way across the garden to see the onion variety Sylvia chose for herself: Southport Red Globe. It grows straight and green, flush with life. "When the little husks around each of the seeds begin to dry out, I will collect them. Of each onion variety I will ask: Does it grow well? Does it store well? Does it taste good? Does it produce seed within one season, with the climate we have here?"

In this experiment Sylvia is a plant selector, not a breeder. Breeding involves producing a specific target. "What I'm doing feels more like a partnership, letting the plant express itself." She talks about phenotype, which means observable traits, and genotype, which is genetic makeup. "Plants will respond in one season to my circumstances. Maybe there's different soil, a different elevation, something slightly different from the place the plant previously grew. Traits hidden because of conditions can surface," she said. "For example, I got a leek from William Woys Weaver. It was originally hardy only to Zone 7. After growing it for a number of years and selecting for seed production the plants that survived winter, it's now winter-hardy here in Vermont. It still looks exactly like the leek I got," she continued. "But I can leave it in the ground all winter. That leek had the genetic capabilities of adjusting to this place."

Sylvia has another idea for an experiment. She wants to grow Sheepnose pepper and allow it to adapt year by year to her central Vermont environment. At the same time, she will save some original Sheepnose pepper seeds for ten years in a freezer and then grow them out. She wants to compare the two peppers. Are they the same? Or has the grown-out pepper slowly and visibly altered itself?

Like seed, each of us has traits hidden deep inside that under the right conditions can emerge. Any of us can be selected and developed. We can become the people we've always wanted to become. We can respond and adjust, sure, but even more important, we can express ourselves. We can become something even stronger and more useful than we were before.

I'd heard Sylvia was working on dehybridizing the Sungold tomato. I've come to talk to her about this. Sungold is a popular tomato—a thin-skinned, orange cherry developed by the Tokita Seed Company of Japan—that I reject in my garden because it is a hybrid, obtainable only by purchase. By growing Sungold every year and selecting for the desired characteristics, Sylvia hoped to have a stable, open-pollinated version after about seven years. These seeds would produce Sungold tomatoes that would be open-source, in the public domain.

"I stopped that project," she says now.

I wait, hand poised over my notebook at a bed of arugula.

"The Sungolds always cracked," she said. "But Tim Peters of Peters Seed and Research developed Sweet Orange II that is totally crack-free. And Plumgold emerged in one of the years I was working. With great tomatoes like that, there was no need for my project to continue." I subsequently learned that other breeders have used a parent line of Sungold in order to create an open-pollinated version of this variety, including Tom Wagner, who bred Flaming Juane and Flamme Burst.

In Sylvia's garden I saw things I'd never seen before: salsify with its purple flowers and also *Scorzonera hispanica*, black salsify or oyster plant, nothing most people would plant in a flower garden but a rage of yellow flowers.

Sylvia turns. "Look at these little flowers with their little green tips. I mean, I ask you . . . " And she doesn't need to finish.

She pauses beside spinach. "Not many people realize," she says, "that it has male and female plants." She shows me the difference, using plants that overwintered under snow. "Once the seeds begin to mature, the male plants die back." This is important to know. I hadn't known it. Yet I had seen spinach plants dying without producing seeds and I had wondered. "The pollen is so fine it's like face powder," Sylvia says.

There is so much to learn, I am thinking.

"One of the fun things I am trying to do here is to garden as if we no longer had oil," Sylvia says.

"Why?" I ask. It is, of course, a rhetorical question.

"Because soon we won't have any!" She laughs, not because she thinks this is funny but as if the line requires some emotion and she can no longer cry about the fact we have reached the zenith of global oil production, and from here on out, we have to learn to live without it, because oil will get more and more scarce, and thus more costly. The laugh is wry.

I am reminded of a Christmas card my friend Susan Murphy received. It said, "Words for a post-petroleum economy: It's been a great party, but it's time to go home! Home to our grandparents' ways: growing our vegetables, traveling less, treasuring home, family, and friends more." That doesn't sound like a Christmas card, but it was.

"You're a hero," I say now to Sylvia.

"Just a gardener," she replies.

She asks if I want to see the seed collection and I do, so we go inside to the basement, where a series of dorm refrigerators are filled with neatly organized bags and packets. Sylvia is cultivating a group of gardeners to be curators of this collection. They meet monthly, share seeds and gardening know-how, and work on collaborative projects to preserve worthy and endangered varieties. Sylvia's little seed bank could replenish a region, start a lot of gardens, feed a lot of people, which is what makes this remarkable woman a revolutionary and an activist, although she might never call herself those things. What she would say is that she's just trying to keep all of us well-fed.

In later correspondence with me, Sylvia eloquently summarized her mission with her garden. "What I'm aiming for is taking the local-food movement to the next logical level," she wrote, "which is to establish a supply of locally grown seed as the underpinnings of a local food supply." As she explained: "Currently the seeds we order through commercial catalogs are grown literally all over the world. They were grown under radically different growing conditions than those that exist where we live. Further, we don't know how the seeds were grown, when they were harvested, or under what conditions they were processed, stored, or transported. It makes perfect sense to grow not just the plants but also the seeds in the area where the food will be consumed, giving them the opportunity to adapt on the deepest level."

Sylvia's wisdom and advice is worth repeating: "The logical next step for the local-food movement is to establish locally grown seeds."

— 9 —

keeping preacher beans alive

THE MUSIC IS LOUD for a nature gala. A curly-haired woman with a guitar, backed by a drummer, is singing "Daddy, Won't You Take Me Back to Muhlenberg County." I'm eating a plate of coleslaw and baked beans, having figured the pulled pork to be industrial meat. The singer is rocking out and people are piling their plates high at the buffet—a fund-raiser for the protection of southeastern Georgia's Satilla River, held at a lodge near Nahunta. It's just me at a long table covered with a white cloth until a man who looks familiar hollers down.

"Mind if I join you?"

"I'd love the company."

"Do you remember me?"

"I remember your face."

"I teach at South Georgia. Doug Tarver is the name."

I ask him how he's been spending the summer and he says his garden is about to kill him.

"If the seed catalogs listed work-hours, nothing would sell." I'm having to yell to be heard.

"It's not that. I have a bad back."

"Not fun," I say.

"But I love it so much I do it anyway."

"You growing any old varieties?" The song ends and my question is too loud.

A spark lights up his face. "I am," he says. "A bean."

"What's the name of it?"

"Preacher bean."

The beans were given to Dr. Tarver's grandmother, Katie Tarver, a devoted gardener who lived in the piney woods of northern Louisiana, in LaSalle Parish. They were presented to her by a country preacher she admired, in the year 1912. The preacher was called to minister elsewhere, and his parting gift to Miss Katie was a handful of tan and purple string bean seeds that the family has kept alive by obeying one important rule.

The preacher asked Miss Katie to make the same promise he had given—to save two year's worth of the best seeds, in case of crop loss. By planting time, unable to remember the name the minister gave the beans, Miss Katie called them Preacher beans. "They grew vigorously and produced exceptionally large quantities of green and purple beans," Dr. Tarver said. Since 1912, for a century, the Tarver family has grown and shared these seeds.

It's a great story, for a party or anytime. Stories like this urge us to make sure our own lives contain such stories that can persist, that can inform and encourage us.

This brings me to another point. An heirloom variety of seed, besides being a genetic resource, has another quality. It is a cultural resource. It has a story. The story changes as time passes. The story Dr. Tarver told about Preacher beans is not the one Katie Tarver told, which is not the one the preacher told. My story now has added another layer. And so the story grows, like humus on a forest floor.

Seeds, as Will Bonsall put it, are poignant and pregnant with story. He told me about a seedsperson and agricultural explorer named Jack Harlan who traveled to Turkey in 1948 to collect plants, especially grains, for the USDA. In Turkey, Harlan discovered incredible diversity, but he didn't record the names of many seeds he found there; they were simply identified by serial numbers upon his return. In other words, Harlan was collecting raw germplasm. "So you don't get information about utilization," Bonsall had said. "You don't get culture, tall tales." (Several decades later Harlan returned to Turkey and was astounded at the number of varieties he was unable to find.)

In her work, anthropologist Virginia Nazarea, author of *Heirloom Seeds and Their Keepers: Marginality and Memory in the Conservation of Biological Diversity*, writes about the relationship between agrobiodiversity and cultural memory. Nazarea's scholarship led her to two wonderful epiphanies: a) that a seed is a cultural resource as well as an agricultural one and b) that despite homogenization, tables do not look the same all across the country. Independent seed savers, according to Nazarea, play a significant role in the conservation of diversity.

Seed savers embody what she calls "marginalities of the mind," a play on Vandana Shiva's *Monocultures of the Mind*, a treatise in which Shiva posits that dominant bodies of knowledge, usually developed by economic forces, create monocultures of thought by squeezing out local alternative systems. Nazarea says that seed savers occupy the margins, hence the term "marginalities of the mind"—defying the homogenization of industrial agriculture in a "playful resistance" or "resistance of the weak." Seed savers are "not burdened by the rebel's ire but rather moved by a searching, creative spirit," she writes, resulting in the persistence of both cultural and genetic diversity. Nazarea calls on growers to "wean ourselves from the historically colonial appropriation of plant genetic resources."

The work of seed savers counters loss of memory, identity, and sense of place—and this is especially true for those immigrant gardeners who bring their own food crops with them. Nazarea advocates a practice called "memory banking," parallel to seed banking, which preserves cultural information alongside genetic and agronomic information. This will, as she writes, "make sure that biodiversity is not decontextualized or divested of emotional meaning and cultural significance." Saving seed and saving germplasm are hugely different beasts.

Seeds are multipronged. They have so many pouches, full of stories. A seed is a city full of avenues, a forest traversed by trails. What happens in the garden, in the field, in the kitchen, in the laboratory, in the warehouse, in the store, in rural communities across America? What happens?

As I think about seeds, their names and stories, an image that keeps coming to me is a bird, a swallow, flying over the world. Tucked inside the feathers of its wings are thousands of tiny seeds, daisies and asters and clovers and more kinds of grasses than I will ever learn to differentiate. Each is a story, and it grows.

Every year the Seed Savers Exchange publishes an annual handbook of their members' seed listings, called the *Seed Savers Exchange Year-book*. In it, seed savers are identified by code—their state abbreviation, followed by the first two letters of their surname, followed by their first initial. I am GA RA J. In their listings, seed savers are encouraged to offer cultural as well as agronomic information about the varieties.

Once at a gathering of seed savers I asked the seedsperson John Swenson a burning genetics question, basically the same one I asked Will Bonsall. If you test the genetics of all the listings in the Seed Savers Exchange, the *Yearbook* would be much smaller, maybe by half. So why save all these varieties? Is seed saving as much a cultural enterprise as a genetic one?

"Your use of the word cultural is significant," Swenson said. "We're preserving a cultural heritage. Wild Goose bean came from the crop of a goose somebody shot. Is it something else, another variety? Maybe it is, maybe not."

"If so much energy and effort is being put into the maintenance of varieties," I said, "to winnow out duplicates would make sense, it seems."

Swenson looked at me as if I were a six-year-old. "From the point of view of a seed saver, genetics is almost insignificant," he said. "All these stories, these recipes—that's what matters. Take garlic. Genetically there are maybe thirty genotypes. But you have hundreds of names floating around."

"So we're supporting the preservation of interesting names?"

"Each name has a story. Each story has a purpose. We're supporting the preservation of human culture."

I had to digest that. I love story. But the scientist in me wants to be efficient.

Swenson's reference to garlic hit close to home. I am one of the people saving a garlic because of a story. A woman who believes in marriage gave it to me. A few years ago, I was in charge of a seed exchange at a local-food barbecue put on by the Okravores, our group of local-food producers and eaters in southern Georgia (with a listserv and a Facebook page). The barbecue was happening at the Agrirama, a living history museum in Tifton, Georgia, and I was standing behind a picnic table crowded with jars, bags, and packets, as well as heaps of dried sunflower heads and brittle bean pods.

At such booths those folks disinterested in plants quickly weed themselves out. A cursory glance over a table filled with containers of seeds will inform such people that nothing of interest is to be found. They will proceed to the barbecue line, where every single mouthful of food they eat is absolutely dependent on a seed. In my perverted way, I have begun to think of the ones who walk on as the sinners, the lost souls.

The others—the lovers, the botanists, the landscape architects, the farmers, the poets, the ecologists, the home-canners, the preachers, the children, the old people—they will wax sentimental over seed heads.

At the cookout of which I speak, a mature woman with her hair fixed (with her hair barbecued, Silas said when he was little) paused at the seed table and looked at everything. She floated there so long I knew her for a saint.

When finally the airwaves were clear, she said, "What are these?"

"Old-timey kinds of seeds."

"That's what I thought. I love this."

"Great," I smiled. "Do you garden?"

"All my life," she said. "I have an old garlic I want to tell you about."

"Oh, yes? What kind?"

"I don't know. We call it Marriage garlic."

"Hmmmm."

"My grandmother gave it to me when I married and moved out on my own. She told me that as long as I grew that garlic, my marriage would last."

"Could you share a start with me?" I asked. "My marriage is in good shape, but I'd sure like to grow it."

"I'd be happy to."

A few months later I received a package from a Jane Howell. "I received this garlic from my grandmother when my husband and I married," a letter reiterated. "I was told she had gotten it from her grandmother with the story that as long as you grew this garlic, your marriage would last. Her marriage lasted over sixty years, and Raymond and I have been married forty-five years this year!"

The garlic made my mailbox smell like an Italian restaurant.

A few months passed before I could plant my garlic. We were in the process of closing on the farm, and I didn't want to abandon the cloves in a place I knew I was leaving. Just after we moved to our new farm, I located the few cloves Jane had mailed. They had almost given up the ghost, but I shucked them into a hole in an herb bed beside the kitchen. One still contained a germ of life and sent up a strong green spear to spice up our marriage kitchen. It's on its third year at my house and I hope it lasts for many more to come.

In the middle of the week following the benefit where I sat beside Dr. Tarver, who told me about his family's Preacher beans, our letter carrier honked in the front yard. The package she delivered contained a sackful of beans. Dr. Tarver had written out the story. He had remembered the preacher's second rule: "You are to share these seeds. However, if someone does not save his own seeds, you are not obligated to give him any the second time he begs."

"From time to time I thumb through seed catalogs," he wrote, "hoping to identify the real name of these beans. If the truth be known, I hope I never do. They are Preacher beans."

The season being late, and with an empty space in the field, I soaked half of my gift immediately. The next day I sowed them, and they germinated as quickly as any seed I've ever planted.

My garden brims with storied varieties, plants linked to anecdote and legend. Whippoorwill Field pea, mentioned by Edmund Ruffin (the Virginian fabled to have fired the first shot of the Civil War) in his 1855 homage to cowpeas, surely is named because its light speckles resembles a whip-poor-will's eggs. Aunt Ruby's German Green tomato came from Ruby Arnold of Greeneville, Tennessee, who passed away in 1997. Old Time Tennessee muskmelon doesn't ship well but is supposedly so fragrant it can be found in the dark. Amish Paste tomato came to me from Lancaster, Pennsylvania, via a seed saver in Wisconsin and represents one of many vegetables stewarded by the Amish.

My garden contains one large bed of Preacher beans! I have extra laid by. The plants act and look an awful lot like Rattlesnake beans, but they'll always be Preacher beans to Dr. Tarver, and they'll always be Preacher beans to me.

oakreez

To APPRECIATE THIS STORY, I have to describe my friend Jack Daniel. He's a tall guy about fifty who used to be a high-rolling building contractor in Savannah. "I really think that working most of my life for the extremely rich and unhappy is what caused me to turn from construction. This fueled my exodus." He moved out to the country, to a ranch-style home on family property near Surrency, Georgia, where he started making wine and hunting deer and growing a garden. Mostly he loved to ride around the woods on his four-wheeler. "In our woods I felt and still feel like the richest man on earth," he said.

Raven and I became friends with Jack when he took on the job of ousting a local superintendent (also his uncle) who had misappropriated some money. It was an unlikely but fun alliance, since Jack was much more conservative, oddly so. Jack called us collectively the "Cavalcade of Fools," although we managed to bring justice to the school board.

One thing about Jack, his name fit him. And it helped him become a great storyteller. Of course, to be a great storyteller you need experience and Jack had it. He'd lived a crazy life of motorcycles, parties, and a few good women. He rubbed elbows with the elite. Somehow he got himself in jail overnight with a cellmate named Genesis and the bailiff cranked down the thermostat. "I've been in here before," Genesis told Jack. "And they ain't ever left the lights on all night nor has it ever been this damned cold!" To Jack, everything was a story and a funny one. Most of our conversations concerned local politics, so I was surprised one day to get a different kind of letter from Jack.

"I heard you were working on a book about seeds," Jack e-mailed me. "I have toted around these okra seeds for the past fifteen years or so . . . no shit! They've been sunk in an ice chest, frozen, unfrozen, moved, carted from here to there and to and fro, and I've finally planted them and damned if they ain't sprouted! Most of the seeds even had little sprouts that had died coming out the end of them. So, not being too optimistic that they would grow when I finally got around to planting them, I planted them heavy in a short row and every one of them have come up! So, now I need to thin them."

He went on to tell the story of how he had obtained the okra, which was described to him as Longhorn okra, in coastal Georgia. "Back in about 1994," he wrote, "I was traveling daily to Long County, about ten miles east of Ludowici, to do a job that I was working on there. Each and every day I passed this little farm. I noticed this old man and lady out doing something in the field in the early morning hours as I passed. Finally, I couldn't take it anymore, wondering what it was they were doing and I stopped one morning."

"Just like always, the old man and lady were out in this fairly wet field that I'd been studying and like always they each had on these giant straw hats, long sleeves, and a five-gallon bucket as they worked the rows, traveling parallel, working two rows at a time. The plants resembled a plant I had in my garden, but the stalks were like cornstalks and the leaves like a giant hibiscus tree. After getting their attention, I questioned them as to what it was that they were picking and the old fellow told me, 'These here is oakreez.'

"Looking into his bucket, I saw something that I'd never seen before and in it were these 'oakreez' that were nearly as tall as the bucket and stacked standing up! The old guy ended up giving me some of the seeds from his freezer and I've carried them around ever since. I also was given a demonstration involving how to know if an 'oakree' was a 'good un' or a 'bad un.' It was the old break-off-the-tip test and these huge things passed! He also took out his knife and cut one and it cut like a regular three-incher.

"Anyway, I do have these things and if you'd like some of the plants, if you've got a good wet area, they should grow until the first frost, producing thousands of seeds and maybe a meal or two for y'all."

I got the plant in 2009 and I've grown it for three years now. I call it Long County Longhorn okra. The proper name for it is Cowhorn okra and it grows about three times as large as regular okra. It gets about twelve feet tall. Jack was right, it's closer to a tree. I'm sure the old couple are dead by now, and I like to think of them from time to time as I weed around the okra trees—they whom I never knew but who gave me, through Jack, a wonderful gift.

— 12 —

the poet who saved seed

I HAD ONE CHANCE to meet Jeff Bickart and luckily I took it. A notice on a bulletin board at the general store in Craftsbury, Vermont, led me to him.

FREE SEED
11 varieties of beans
18 varieties of potatoes

I had been teaching for a week at Wildbranch Writing Workshop at Sterling College nearby. Jeff gave me directions over the phone and I arrived at a modern homestead on the Wildbranch River, a tributary of the Lamoille. A tall man was weeding and mulching grapevines in expansive gardens below a timber-frame house when we arrived. He looked to be in his late forties and wore a sage-colored fishing shirt and dark workpants. A watch dangled from a belt loop.

"We're the people who called."

"Welcome," he said. He looked in no manner ill.

"You have a beautiful place."

"Yes," he said. "We're very happy to be here."

We stood in the cool Vermont afternoon looking down into the floodplain meadows toward the Wildbranch. The architecture of the meadow was wildflowers.

I had brought a young writer, Holli Cedarholm, with me, because from her writing I knew she was interested in agriculture and because

I liked her. Holli was a quiet, rosy-cheeked, down-to-earth redhead who exhibited a rare inner strength. She was camping that week to save money, whereas all the other participants were in the dorm. My husband Raven made three of us.

In the eight years almost to the day that Jeff and his family (high-school Spanish teacher wife and two children, seven and nine) had lived on their eighty-seven-acre farm, they had been planting—160 gooseberries, 45 apples, blueberries, pears. Gardens surrounded us, and beyond the beds overflowing with verdure, green pastures trailed to the river. Jeff's was a storybook home, made more picturesque by the fervor of Vermont summer. I was smitten, not to mention envious and inspired, as were Holli and Raven; Jeff, poor man, was bombarded by questions.

"Do you grow for market?"

"No, mostly for the family, for subsistence, and the extra we sell."

"Do you have animals?"

"I'm much more interested in fruits and vegetables."

"Do you work off the farm?"

"I'm lucky. No." He swept his hand across his light-brown hair.

"What brought you to Craftsbury?"

"I came to teach at Sterling College and was there from '98 to 2001."

I asked what he taught and he told me: ornithology, botany, organic gardening. "Well, you want to go see some seeds?" Jeff asked.

Inside his home the living room was bright, cozy, and clean, its pine floor topped with a red braided rug next to a cranberry sofa. Large windows let in plenty of light. The floor plan was open, and we could see across to the kitchen and the table that served as the dining room. Every element of the home was cheerful, carefully chosen (mostly of natural material), and in its place.

Hanging beside the door was a noteboard with the line, ALWAYS BE FINISHING SOMETHING. Someone had jotted below it a list of jobs—START CABBAGE and TOPSOIL.

Something more was in the room. There was an unnatural undercurrent, an aspect of sadness—or was it contemplation? Was it because all light came from natural sources, the wide windows, that the house seemed to be waiting—waiting for the children to arrive on the school bus and also waiting for something more?

Jeff led us past a spinning wheel (which we learned belonged to him, not his wife—he was a weaver and a knitter), past a blond pine table on which lay a copy of the latest *Small Farmer's Journal* and Mother Earth's book on solar food drying. We passed into an office lined with bookshelves, where a computer's screensaver twirled random geometric shapes at a desk. Jeff knelt on the floor and began to unpack a few boxes full of quart-sized canning jars, neatly labeled on each lid. The jars rattled like shakers.

"Mostly I grow easy seeds, ones you don't have to bag," Jeff said. His eyes were pale blue behind square-rimmed glasses. His hair was brown, thin.

The jars were filled with beans. Odd-looking beans.

"Take a look," Jeff said. "You're welcome to a start of whatever you want." Holli's hands ventured among the jars and she ceremoniously picked up one. I saw something of myself in her. I chose one whose label read BEAN/BUSH/DRY: AGATE PINTO, followed by a series of abbreviations that indicated where Jeff obtained the seed. The Agate is a cultivar of the pinto, mottled and slightly flattened, bred to be a bush, although it still throws a few runners. It is not an heirloom per se, but an open-pollinated variety created by Rogers Brothers Seed Company (bought by Sandoz in 1975, which merged with Ciba to form Novartis in 1996, and which became Syngenta in 2000).

I liked the way Jeff's beans rolled and clattered in their jars when I rotated them in the light. Some varieties were breathtaking. Calypso—a half-black and half-white bean, with tiny black dots in the white area— reminded me of the yin-yang symbol and of orcas. King of the Early was kidney-maroon mottled with tan. There were Black Turtle beans, Pawnee beans, tepary beans, Cranberry beans—all stunningly patterned and colored. "I went through the Seed Savers Exchange catalog and picked the prettiest," Jeff said.

Practical Holli asked which ones tasted good but Jeff hadn't sampled all of them, he said. Seeds were his first priority. This year he hoped to have beans left over, after planting and after sharing, and those he planned to eat. One day soon he'd like to be producing a hundred pounds of beans a year.

"My greatest interest is in life-sustaining crops," he said. "Grains, beans, onions, stuff you can really live on, that provide protein and carbohydrates, calories. Crops that are fundamental to keeping you alive." Much later I would revisit this conversation in memory and realize the poignancy, wistfulness, and doggedness in Jeff's words.

"Last year, in 2007, I began growing barley, seven varieties, including a Purple Hullless, which does not require threshing. Now I grow oats, millet, and wheat for grain production. I'm also interested in the home-scale growing and processing of oil seeds."

"Such as?" I asked.

"Safflower, sunflower, flax."

Eager to share his beans, Jeff produced small plastic bags and I chose Paint; Tiger Eye; and Mitla Black—a tepary, which is a prehistoric, drought-resistant bean, this one from Oaxaca's Mitla Valley. I was careful to curb my appetite, knowing that heirloom seeds are like sourdough culture. They're pets. You may not have to feed them twice a day, but they come with their own needs, and once you assume them, you're responsible. You have to do what needs to be done to take care of them. I think Holli accepted only one variety. Her garden is small.

One of Jeff's jars was full of beans that didn't look as if they belonged together, some solid, some speckled—MULL KIDNEY bean. He had ordered them from Will Bonsall.

"These are supposed to be red," he said. "When I grew them out, this is what I got. Both white and brownish-gray speckled beans contaminated the red. I don't know if these crossed in Will's garden or in mine. I'm going to grow them out and see. I may have a new variety here."

"If you do, what will you name it?"

He smiled shyly. "I don't know."

Beyond Jeff were full bookshelves. Both editions of Suzanne Ashworth's classic book *Seed to Seed* stood at eye level. I saw novels, volumes of nature writing, collections of poetry. Because the computer waited at the ready, and because something was eerily familiar about the quality of silence in the house, beyond our chatter and the jangle of seeds, I had a hunch.

"Are you a writer?"

He hesitated. "I attempt to be."

"What genre?"

"Poetry."

Of course. And I thought, "You don't have to be a poet to save seeds. But there's a good chance that you are." I wanted to ask him for a poem or two, but he was already giving us a lot. So I asked him, instead, how he got interested in seed saving.

"It was soon after my wife, Jennifer, and I married," he said, "when we dug our first garden, in the summer of 1994. We got into it together. It dawned on me that it was time to start growing food. Maybe it was finally being married, suddenly we're making a household and households need food, good food."

Fourteen years later, eight of those years on the farm, he grew twenty-two kinds of beans, five kinds of barley, eight kinds of garlic, and twenty kinds of potatoes. In the 2008 *Seed Savers Exchange Yearbook* he offered fifty-two varieties. He'd only been on the farm for eight years. He had done a lot in eight years.

"Do you ever wish you'd got started earlier?" I asked. I think a lot about that, because I came late to the life I dreamed of, and because when I visit the homes of people who have lived in one place for many years, I see the amazing things that are possible if one chooses to stay put.

"It's not too late," he said. "Although it's later than it should have been."

At the time I didn't realize what he meant.

When I mentioned cowpeas, since they grow so well in our Southeastern climate, Jeff brought out a large grocery sack of seed packets. "I tried to grow cowpeas last year," he said, "but our growing season is too short and they never did well. If you can grow them down where you live in Georgia, you can take all these."

I fiddled with the packets. Red Ripper and Papago and Pennyrile and Purple Hull Pinkeye and Queen Anne Black-eye. I didn't need all these varieties. It was June and the growing season was well underway in Georgia—where would I plant them? I'd have to make more beds, make room. But here, in Jeff's office, they would slowly lose their viability and they would die. They would die because the life inside would slowly flow out and nothing would be able to retrieve it. I didn't want the responsibility of twenty kinds of heirloom cowpeas, but better me than nobody. The beans, I told myself, ought to last at least three years at room temperature without much loss of germination, and they would last longer in the freezer. More than a bit regretfully I accepted all of them.

"I understand," he said. "Since I got sick I've had to choose carefully what I can and can't take on."

"Sick?"

"I was diagnosed with cancer. Melanoma. I'm now in remission." That, I realized, explained the trace of sadness in the air and the quietness, the purpose-driven home. Illness is a state of being that makes life more precious than it might be to a person unvisited by meditations upon time.

No one could have remained unaltered by the news. The information took me completely off guard. Jeff was so young and vibrant and healthy—as well as obviously talented, brilliant, and committed to serving the world. Being a hospice volunteer and my husband an EMT, I have seen a bit of death; no amount of society with it, or rationality about it, can diffuse the sense of tragedy. Here was a man so vibrant and good, visionary and hardworking, a man with children who needed him, fighting to live, hoping to remain in remission.

But plenty of people survive calamities of this sort.

Jeff glanced out the window, watching for his children, and with that signal we trooped out, pausing to tour more gardens along the terraced hill around the house. Everything was tidy, weedless, well-labeled, green, and growing—Slavic bread-seed poppies, kamut, fifteen varieties of tomatoes. Beyond the gardens the meadow was filled with forget-me-nots. We thanked Jeff and promised to let him know how the beans and peas grew.

A year will pass before I hear news of him.

the anatomy of inflorescence: a quick lesson

WHEN I WAS A KID, there was a ditty certain to send my friends into spasms of giggles: *First comes love, then comes marriage, then comes a baby in a baby carriage.* What exactly happens between marriage and carriage is a great unknown, but over time every kid begins to comprehend one thing for sure—*something* happens. Awkwardly, blindly, and clandestinely, sometimes brutally, the great mysteries of sex are revealed to us.

The same with plant sex. Most of us are in the juvenile state of not knowing about the business of food creation.

No study of seeds can be complete without a quick review of carpology.

Vascular, seed-bearing plants within the kingdom Plantae are divided into two classes, angiosperms and gymnosperms. Angiosperms—dandelions, lilies, tomatoes, oranges, walnuts, peas—produce seeds enclosed in an ovary; gymnosperms—pines, cedars, cypress, spruces, cycads, gingkos—produce seeds on open scales, usually cones. I'm just going to talk about angiosperms here. Not that we don't get food from gymnosperms, we do (pine nuts, for example, from stone pines). For the most part, however, our food comes from angiosperms.

Most angiosperms are flowering plants. Here's their ditty: *First comes a flower, then pollination, and that's how a plant does multiplication.* Practices procreation. Increases population. Knows consummation.

In less silly terms, plants construct flowers for a reason. They want to make more of themselves, to vegetate the world. After the flower is produced, it must be fertilized. A flower can be pollinated in three ways: a) with no outside effort, known as self-pollination, b) by wind, and c) by insect. Those in the last class must use bright colors and unusual fragrances to attract the pollinators they need.

Let's look at a simple flower. Most flowers have a calyx, a set of leaflike sepals, and a corolla, which is the petals taken together. The female reproductive organs consist of an ovary, a vase where the eggs (or ovules) are fabricated; a stigma, a disk that receives pollen during fertilization; and a style, a slender stem or stalk to connect the two. Together the female organs are called the pistil. The male reproductive organs, or stamen, consist of an anther—which produces pollen—and a filament to hold the anther. In general, anthers cluster around the styles.

Here the botany of inflorescence gets more complicated. Some plants have *perfect* flowers, meaning they have both sex organs in one flower (such as peas and lettuce). Other plants have *imperfect* flowers, meaning only male or only female organs in one flower.

In plants with imperfect flowers, sometimes both male and female flowers occur separately on the same plant. These are called monoecious, which is Greek for "one household"—such as cucumber, corn, and figs. Dioecious, or "two households," plants, on the other hand, have male flowers on one plant and female flowers on another—spinach, asparagus, and hemp.

Okay, cucumber is monoecious. One cuke plant has both female and male flowers. This is how you make a baby cucumber. After the woman cucumber marries the man cucumber, a bumblebee goes to visit each of them. He visits the man first. The man invites him to see his towers of gold, and while climbing there, some gold dust sticks to the bumblebee's feet. The bee sits with the man and they get a nice buzz on, then the bumblebee says his goodbye. He needs to keep moving, to visit the woman, whom he finds by following the vine trail straight out the man's door. The woman lives nearby. The bee tracks gold dust right through the woman's doorway, and that causes a little cucumber at the bottom of her pretty yellow flower to start growing.

A Boothby Blonde cuke. Or a Diva. Or a Little Tyke. Or a Straight Eight.

Spinach is dioecious, some plants male and some female. This is how to make a baby spinach. The female spinach wants to have a house of her own. She enjoys her solitude and likes living by herself. The male spinach accepts that and has come to enjoy his own space too. But he loves the woman and he knows they need to make little ones of themselves, for they are only annuals and die after a year. When the woman is ready to make seeds, she telephones the man, and he opens up his flowers, which contain pollen as fine as baby powder. The wind picks up the pollen and carries it over to where the woman lives and flings it on her stigmas. This causes her ovules to begin turning into a little cluster of seeds.

Bloomsdale Long Standing seeds. Or Spartacus. Or Giant Nobel. Or Monnopa.

— 14 —

red earth

SOMETIMES I DREAM a tree birthed me; I came tumbling like an apple out of its limbs. I came to a causeway and looked out across my father's and mother's faces, which were shining in the sun like the Gulf. I saw many beautiful things. I saw love in the eyes of deer. I saw the throats of lilies moving. And I wanted a farm. I wanted a farm at the border of wilderness.

I could not escape the terrible yearning.

Even as I was becoming a nature writer, seeking wildness and spending halcyon days walking through the remaining tracts of longleaf pine flatwoods, I battled a piece of myself that was happiest not in wilderness, but on a farm. I had come to think of a societal continuum that begins with wildness on one end (hunting and gathering for food), moving through agrarianism (settling down and tending a piece of land), then through industrialism (an urban life), into technologism (whatever that lifestyle is). A tract of land could sustain a forest or a farm or a manufacturing plant or a bank of computers operated by robots. If wildness was on the left of the continuum, I wanted all movement in terms of land use to be from right to left, always toward wildness. But though my hope for land is that it tends toward wild, the truth is that I am probably happiest somewhere in the middle. My friend Rick Bass once said to me, "What I would want, after working in the fields, would be to step away from the plow and enter an old forest, where I could walk, and rest at the end of a day of hard work."

Once Silas left for college, every morning soon after I woke the longing accosted me. My mind turned to thoughts of what my life would be like, had I a place of my own. How—if I had land, trees, fields—different my decisions would be. I think perhaps the feeling derived from the idea of *cultus* and the instinct to care for something. I needed something to cultivate after Silas left. Every morning my thoughts arrived ultimately at the same question: Where is this place?

My husband and I actively searched for it. We wrote friends and strangers alike.

> *We are looking for a homestead. We are searching for an old Cracker house on a piece of land preferably in southern Georgia, in the coastal plain. A fixer-upper is good. Over fifty years old and historical is great. We are not looking for a modern-style ranch house or a brick house. We are not looking for highway frontage, but prefer to be on a dirt road. We'd love to have field, forest, a barn, or outbuildings (rundown is okay)—maybe wetlands or a creek or a pond. The more land the better, taking into account that we're not millionaires. We need pasture. The upper limit of our comfort range is about $220,000. If you know or hear of any such place that comes close to this description, please let us know. We'll investigate all possibilities. Thank you so much. Please pass this message along. And best of luck to you in finding the things you're seeking!*

We searched over a year for the right place, hours each week scouring the papers and online Realtor listings, more hours on the phone, more traipsing with Realtors around places that would never work. We put notes in mailboxes outside houses we fancied. You never know when and where you'll find what you desire. But you get tired and you get discouraged. Every morning I knew I had to keep looking, with renewed vigor.

Meanwhile, we had been collecting farm animals and seeds and herbs and skills. All that was striking root. I needed a place to practice.

Part of the urgency Raven and I felt about finding a place was the growing body of evidence substantiating collapse, especially of the climate—and not simply the knowledge but the experience of it. During the two years of our search, the South emerged from a severe, three-year drought. Tornadoes in March were ransacking towns, tearing down schools and neighborhoods, killing people. More and more, the statistics pointed to the need to be settled in a community and able to provide at least some of one's own basic needs. The national (and then international) housing catastrophe, as well as the stock market crash in the fall of 2008, naturally increased our panic.

During the second September of that long search, I had a gut-wrenching dream. A storm was coming. I had been at a gathering of people, many of them friends, and worry had descended on us. It was palpable. I was leaving with Silas when suddenly he and I were standing on the edge of outer space, on the perimeter of the very atmosphere. All around us the biosphere was blue, all shades of blue, swirling, something you might see if you were doing psychedelic mushrooms. I knew the blue meant ice. The blue hues were eddying, dragged around by speeding global winds. We could feel the cold wind all around us, it was a monolithic wind. I felt amazed and also helpless, and I remember thinking that at least Silas and I were together. Where the colors were powder blue I knew the ice had been melting.

That was the catastrophe. But now the ice was beginning to solidify again, and the whirling winds were turning all colors, rainbow colors, bright and vivid. We made offerings, Silas and I, of what we had, which was strips of potatoes. Then a man appeared. I didn't know him and neither did Silas, but the man was holding a baby. That would become the important fact.

That's all of the dream I remember, and maybe that's all there was. I've never studied dream interpretation, but a couple of friends have, and I know that it was both a dream of warning and one of hope, a millennia encapsulated, a collapse and a rebuilding. There was hope in the colors and in the baby.

One day, jogging my daily two miles, I knew in my bones that the place for which we searched was available finally, that we simply had to find it. Such a premonition does little good when you have been looking already for so long, when you have a vast area in which to search (Raven and I had expanded our search into the Carolinas, Virginia, and north Florida, although we were still focused on south Georgia and the most rural parts of it). The last thing we wanted as the news worsened was to blast carbon dioxide into the air while riding on dirt roads, looking for a home. We were, at the time, living on family land and could have remained, but there we owned nothing. We were required to ask permission of my family in order to make changes that would simplify our existence, allowing us to be healthier and safer and more self-reliant, but many of our requests were denied. Our vision was not the vision of my family. My father, a junkman, continued to haul in load after load of material detritus and immaterial valuables.

Our farm came on the market when its owner, a developer, was unable to liquidate properties elsewhere. The FDIC forced the owner's banks (he had a first and a second mortgage) to put pressure on him to sell. He wasn't eager to part with the place.

One day I saw a picture of the house online. I remember the afternoon well. Raven was particularly dejected about our search that day, stymied as he was in his ability to move forward with his vision. I was on a writing deadline and needed to pack for a speaking engagement, but I'd sneaked a look at a Realtor site.

"Why don't you ride over and look at this place?" I asked Raven. "Just to make sure. I know it's not where we want to be, but it looks interesting."

He was back in a hurry. "Drop everything," he said. "We've found our farm."

The house was two stories and painted white with forest green trim. Its metal roof was forest green. It was built in 1850 by Lawrence Pearson using native longleaf pine in the Federal style, although during renovations the front porch had been wrapped around in the style of Victorians. The wood for the house's construction had come from surrounding flatwoods and had been milled less than a mile away, at a mill operated by Pearson's brother.

The house sat on 46 acres. To the north and south were large pastures. To the east stood a mature pecan orchard where wild onions scented the fall air, and beyond that ran a dirt road. To the west mature deciduous forest descended slowly to a cypress-lined, blackwater, lilting stream named Slaughter Creek (because of a battle between Native Americans and white settlers). From any window of the house, only nature was visible—no neighbors, no streets, no electric lines, no gutters.

The farm was located in the delta of two rivers, each of them a couple miles away. To the east, the pristine Ohoopee carved its way through white sand, joining the fat Altamaha south of the farm, which drifted lazily toward Darien on the coast, some eighty miles away. The yard of the house was planted with redbuds, cedar, holly, and crepe myrtle trees. Best of all, in front of the house grew an ancient swamp chestnut oak—a landmark, tall and awesome—which covers an eighth of an acre by itself and which was dropping incredibly large acorns, its seeds everywhere.

In the fall of 2009, I would remember all the months of waiting; all those harangued months filled with longing for a place of my own, the place I dreamed of, where we could live the life we desired, where we could build things that would stay, where we could stay, even where we could be buried; and all those months of evenings when I searched newspapers and magazines and websites for the one ad that would call out to us, a home that we could afford. I remembered all that one morning teaching a nature-writing class as a visiting writer at Muhlenberg College in Pennsylvania, while wind made the yellow trees sound like rain and somewhere a tractor puttered, while back home in southern Georgia my husband was in a lawyer's office, signing for both of us, and although I wouldn't be able to move in until I returned home a few days later, I would soon have the farm I dreamed of.

Consider the possibility that I had been moving toward this land all my life. Consider that it was meant for me.

No one ever settled in more rapidly. We unpacked boxes in the evenings, and during daylight hours erected a mailbox, set up the chicken tractors in the meadow, planted fruit trees and berries, got a garden spot harrowed, planted more shade trees, fenced two pastures, cleaned out and moved things into the barn, built a hog pen. One day the name of the farm came to Raven: Red Earth. After five months we had a goat shed and a sheep hut. We'd thrown a few parties and entertained a constant stream of visitors. Our EGGS FOR SALE sign was up.

There were all the usual chores, egg collecting and supper making, combined with an endless list of tasks for setting up the infrastructure for a new farm to sustain us as quickly and beautifully as possible. We began to haul in loads of biologic debris, manure and spoiled hay and wood chips, cardboard for mulch, shrimp heads from the fish market, ash from a neighbor's outdoor furnace. We needed meat for the year and so Raven was up early many mornings before dawn, finding his way to a deer stand he'd erected at the edge of the woods. Sometimes I rose and layered my warmest clothes and went in another direction, to a tree stand the previous owner had left at the northwest corner of the property. In the mornings the roadbed would be scribbled with hoofprints and sometimes we would see deer in our headlights when we came home late from a reading.

That first winter, I was eager to get my thirteenth and, I hoped, last garden planted. I remembered all the gardens I had created and left behind—because I grew up, because I broke up, because the rent got raised, because I went off to graduate school, because graduate school ended, because, because, because. In a stable home, with no intention of ever moving, I could curate the seeds I had collected along the way, had read about, or was saving already as best I could. I could guard seeds more securely. We began to double-dig beds four feet wide and ten feet long, fifteen in one garden and twenty-five in another. As I dug I thought about a line from author Fred Magdoff, who said there are three sources of soil fertility—"the living, the dead, and the very dead." He meant microbial life, decaying organic matter, and finally humus.

I spent exciting evenings speculating over seed catalogs and finally made my orders, all open-pollinated varieties. When the seeds arrived, Christmas twice, I scrounged up every seed I had stored in the freezer, in jars and coffee cans and packets, and I organized them by dividing them into crops. All the lettuce packets went together in a bag, all the flowers, all the cowpeas, all the Swiss chard. I put these bags into two larger ones, summer crops and winter crops. In the last weeks of late winter I began to plant.

I almost never write about the hours I spend in the garden, not even in my journal. Something happens to me when I garden. I am fully, reliably, blissfully present to who I am and where I am in that moment. I am an animal with a hundred different senses and all of them are switched on.

I mix planting soil, stirring like a witch at a cauldron of compost, peat, and ashes. Nothing is scientific, nothing is tested. Everything I do is an experiment. I am planting seeds in eight-inch pots, carefully spacing the little pillars of life, little buckets, little germs of ideas. A germ, from Latin *germen*, meaning "sprout," is the seed of disease, yes. But it is also the bone marrow, the pith, the essence, the oil. Night is coming on and I am working quickly. I hear a bird call from the sandspur field across the road. It is the first chuck-will's-widow arrived from the South.

What do a seed and a pebble have in common? Both tend to be small, and rounded, and smooth, and hard. One is harder than the other, stronger than the enamel of human teeth; the other mostly capable of being crushed by enamel—although I know people, myself included, who have broken their teeth on seeds. Popcorn, specifically. Pebbles and seeds can be mistaken for each other, until you drop something in a hole that will never sprout. I think of all the jars of soup beans I have sorted and in which I found stones masquerading. Of seeds and pebbles, each has many kinds, people collect each one, museums are devoted to each, each has a science. And each has a mystery beyond science, the life of a stone unknown, the life of a seed waiting in its patient dormancy. They come in all sizes, colors, forms. One carries information between generations, the other between geologic periods. One deconstructs from the inside out, the other from the outside in: if a pebble evolves it does so because of erosion. Both have clocks, one biologic, one geologic; one quick, one slow. Both lie for a time in the ground.

I am cutting labels from old pie tins, pressing a dried-up ballpoint into them to write the labels. KALE, LACINATO. KALE, DWARF BLUE CURLED. KALE, RED RUSSIAN. KALE, GREENPEACE. I am hoping that the labels won't get lost or become unreadable. Someone told me that old metal blind slats would be good for this. I wonder where I can find some.

It is dark and I am watering the pots of seeds.

It is morning and I am watering them.

Four weeks later and I am transplanting into the long perfect beds, kneeling in a thick mulch of wheat straw, over a layer of wood chips dumped by the electric company because we sometimes tip the workers, over a layer of cardboard. I am watering the seedlings that so shakily stand in their little water saucers. I am sticking more labels in soft ground. I am drawing a grid of the South Garden on graph paper, another of the North Garden, and I am naming what is there.

I am racing nightfall to get three-inch melon starts in the Round Garden, where we concentrate vining crops. I am mulching the seedlings. I am watching them grow. I am side-dressing with aged manure we hauled in from the livestock sale barn, sprinkling worm castings. I am weeding. I have my eye on the first sugar snap pea. I am picking the first lettuce.

Then it is summer and the pigeon peas that my peanut pathologist friend Albert Culbreath gave me are taller than my head. They are eight feet tall, flowering way up there in the most astonishing and giddy scarlet. I have risen from weeding a few stray plants, pigweed and purslane and a chinaberry sprout, from beneath the forest of pigeon pea stalks. I hear the distinctive twittering of a hummingbird. It is a rubythroat, feeding at the pulse flowers above me and making more noise than I think necessary.

I stand still. The hummingbird flies away and lands on the hogwire fence that surrounds the garden. Then I notice a smaller hummingbird a foot away, on the same fence. A good deal of animated hummingbird conversation is underway at the fence; another bird zips from a pecan tree nearby and into the flowers. My face is a daylight moon, staring up through composite leaves, below a field of flowers. I am two feet from a couple of the birds, which are careless, and I realize that they are fledglings, they are leaving a nest nearby, they appear to be learning from their parents how to suckle nectar. They are very excited. For a long time I do not move, they are not startled, and I watch the two young and their parents darting in awkward, helter-skelter ways between blossoms, but unable to hover for long, they careen back to the enclosure wire to rest. I watch until they all fly away through the blue-swirling biosphere.

What I am saying is that lovely, whimsical, and soulful things happen in a garden, leaving a gardener giddy. I am on one side of a row of Striped Roman tomatoes and my son Silas, out of college for the summer, is on the other. He works faster than I, grabbing handfuls of weeds and ripping them out. Sometimes he gets the roots and sometimes he doesn't. I am more painstaking. I oust every blade of grass, every henbit and celandine, grasping them at ground level to satisfy myself with the sound of uprooting, a tearing muffled by dirt. I'm in high spirits to be working with Silas. He's talking to me about his life. He and a friend are on the outs, over a skateboard that one borrowed from the other. I listen, sometimes asking a question for clarity, keeping my head down, pulling weeds, trying not to grab a raspberry cane by mistake.

"Is the fight really about the skateboard or is it about something else?" I ask Silas.

"He thinks I'm cocky," Silas says. All the while the sun is beating down hard and I am wishing for a hat. Raven in the distance photographs flowers.

"You know what kind of tomato this is?" I ask Silas.

"Nope."

"Striped Roman."

Silas doesn't answer.

"It was developed by a man I know, John Swenson. It's really pretty, long and pointy with orange stripes."

"Right on," he says and no more. We finish the row and move to another of younger plants. "I read your emails to Dad," he says, "and I agree with you." He's speaking of a series of emails in which I am discussing college finances with Silas's father, opposing student loans. I'm glad Silas wants to talk, even if he doesn't want to talk about what I want to talk about.

"Good has come of it," I say.

"I'm just saying I thought you were right," he says. "I'm on your side."

"I expect you to always be on my side, when all's said and done," I say. "Because I'll be siding with you. I'll be standing beside you."

"I've always been on your side," he says. We move to the eggplant and fall into silence.

I am alone in the garden, as I most often am, planting more seeds. I am transplanting, thinking of supper, because it's time to cook. I am grabbling sweet potatoes, pulling radishes, stripping basil leaves.

When I harvest the food I eat, I stop to consider the coming seeds as a crop. Some of this thinking is second nature. I can't pull all the radishes. Nor do I want my okra over time to become late okra, so I leave a pod or two of the very first okra on each plant before I begin harvesting. I know some people say not to mix eating and seed saving and those people would eat *no* fruit from this okra, but people have been doing both for centuries. I may be wrong but that feels natural. I am tying cotton strips torn one inch wide around the stems of chosen pods to remind myself and Raven not to pick them. I am gathering cowpea pods dry on the vine along with the green ones; I eat the green and save the dry. I am stuffing the ripe seed heads of parsnips into old feed bags. I am squeezing the pulp of marble-sized Matt's Sweet Wild Cherry tomatoes one by one into a canning jar.

I am making pesto, roasting zucchini, rinsing lettuce, chopping a cabbage into slaw. I am scrambling yard eggs with Swiss chard.

For a gardener and a seed saver there is almost never an evening without handiwork—beans to shell, seeds to flail or winnow, squash seeds to spread out on newspapers to dry, flower heads to rub apart. There are seeds to measure and pack into envelopes, labels to be attached, poems to print out and slip inside.

What do a seed and a planet have in common? Both are rounded, smooth, and multicolored, they hold life, they are alive—a cosmology of seeds, each a heavenly body, sailing through the sky, except all the seed-planet wants to do is find a patch of water, a cloud, and then start to germinate, and keep sailing, looking for a patch of ground, which is simply a planet. Imagine Venus turned into a morning glory seed, lying in a packet in its chill, and how, when you finally plant the strange jewel, it becomes a bedazzling flower with three moons of its own. Imagine saving a world by saving its seeds.

I've had to leave the farm briefly to do another reading at another university. It's done and I'm on a train, ever nearer our home. Tomorrow I will arrive there. I have just finished *Bringing It to the Table*, the anthology of Wendell Berry's work on farming and food, which proves that he was the kingpin of the organic ag movement and prince of the local-food revolution. Always his writing settles me into a deep peace and a hope that my yearning for farm life, for a disappearing culture, can be satisfied. I daydream that I can after all become a farmer, that I can raise a couple of oxen, that our farm can be bountiful. I look at my planner and I see that many days ahead of me for what I know to be weeks are outlined already—the taxes to be done, a book edited its final time, an essay to be written.

And then I am sitting in my garden again, in twenty-first-century America, watching the hens stretching their necks through the fence to eat the Sugar Snap peas. A Carolina wren lands nearby and snatches a wisp of straw mulch before flying away. Overhead, a blue-swirling sky begins to fill with the bright rose and peach of sunset.

— 15 —

pilgrimage to mecca

I WAS IN TROUBLE.

Over twenty years had passed since I heard of the Seed Savers Exchange and now I wanted to go to the annual campout at their headquarters. The trouble was, because of the climate crisis, I had quit flying. I live in Georgia; the Seed Savers Exchange is in Iowa.

My last flight was in April of 2008, from Chicago, where I was stuck by some weather incident or another, to Oxford, Mississippi, for a lecture at the University of Mississippi to herald a new sustainability major there. I am not saying that I will never fly again. If there's a crisis with someone I love and I need to arrive quickly someplace, I will fly. For now, I let the record speak for itself. I quit flying more than four years ago and have not stepped on a plane since.

I studied wistfully the invitation to the campout. I found Decorah on a map. It was far away. So far, however, nothing had stopped me from traveling, since there are plenty of other conveyances besides planes, including feet. I plotted my journey.

I packed a backpack, kissed my husband goodbye—a long kiss— and drove four hours to Atlanta. Somewhere near the fall line the pickup's odometer reached the numerologically powerful 234,235. In a free parking garage at a Marta station I parked the truck, exchanged money for tokens, and boarded the subway, where I sat next to a woman carrying a huge bouquet of carnations and in front of a man who planned to start his yoga practice the following weekend. The subway delivered me to the bus station.

The architecturally artless station was of an era, its floor concrete, its ceiling too low, and its lights too bright. The terminal was packed with poor students sitting on baggage reading Foucault and Faulkner, migrant workers, and other wayfarers, some with children, headed toward new dreams of happiness. A woman at the ticket counter told me that seats were first-come, first-served, unless I wanted to pay an extra five dollars to get a window seat. Determined to get one free, I queued up at the Chicago gate although the bus wouldn't leave for a couple of hours.

When we pulled out of Atlanta, the driver—Miz Off-the-Chain, as she called herself—let us know to turn off cell phones and not bother our neighbors by talking loudly. Keep the bathroom clean, don't get up and walk around, she said. If the bus stops, don't get off unless it is a designated rest stop. At the designated rest stops, if you aren't back at the bus in the time allotted, she would, by golly, leave you. Test her.

The bus crawled north out of Atlanta. It was an older-model Greyhound, with its seats too close together and uncomfortable, upholstered in royal blue velour printed with light-blue greyhounds. THERE'S A REASON WE'RE NOT NAMED AFTER A SLOTH, read a small placard. My seatmate, Brian, was a trucker trying to get home for a wedding. We talked awhile, then I stared out my window. A young Maori woman behind me was a student at Vanderbilt. Her wife was pregnant, she said, due in September, and they planned to name the child Rhythm. I read *The Long Emergency* to pass the time. I slept and woke. I ate a sandwich I'd packed. Through Tennessee and Kentucky we motored.

Around midnight the bus paused for a half-hour layover in Louisville. I lent a hand to a scarred, liver-eyed drifter trying to get his sign off the bus: I AM HUNGRY. I AM STRANDED. I'D RATHER BEG THAN STEAL. In the walk to the station he told me that he sets up at intersections. "You have to be near some businesses," he said. In the vaulted station with handsome wooden benches, I noticed a few travelers dressed in spotless khaki pants and white undershirts still creased from the package, with cheap high-top sneakers. That was strange, so many travelers dressed in those creased undershirts.

Brian was at the vending machines. "Why are those guys dressed like that?" I asked him. "Is that a uniform?"

"No telling. Maybe they work here?"

"I think they're just out of prison," I said.

He looked again and I saw a complex emotion cross his dark eyes.

At every stop through the long night and into the next day, until we arrived in Chicago, I would see men in new clothes killing time in bus stations. Once I saw a few such men at a station with a guard, waiting to board my bus.

"So that's what you get when you leave prison?" I asked Brian. "A set of clothes and a bus ticket?"

"Surely they give them some money, especially if they worked."

"Imagine having to start all over. I wonder if they have a place to go, if somebody is waiting."

"I bet so."

I pulled small bills, fives and tens, from my purse and stashed them within easy reach, should an occasion arise to dispense them.

By sunrise, the bus was running a couple hours late. As we traveled through the vast corn-country of Indiana on a cramped bus, I thought about the glamorous seed-collecting expeditions—the Bartrams in the South, Jamaica Kincaid to China and Nepal, Gary Nabhan to Tajikistan. I wander in my varied wardrobe, passing as nobody.

In Chicago, close to noon, I said goodbye to my new friend Brian (whose ride would be waiting), shouldered my pack, and walked five blocks to Union Station, which has become familiar to me during the past few plane-free years. I ordered a burrito at my favorite stand. As I ate I heard my name over the intercom system telling me to go to a certain counter. "Is this your ticket?" the clerk asked, and it was. I'd dropped my ticket and a worker had found it and turned it in. "She could have turned this in for money," the clerk said. "And you would have had to buy another."

"People are basically good," I said. "Aren't they?"

"Yes," she said.

I waited for the 2:15 p.m. departure of the train, the Empire Builder. Now I had been traveling twenty-four hours.

Once on the train, I was more comfortable. I had last ridden this route to deliver a talk in Winona, Minnesota, along the Ice Age National Scenic Trail, and the stops were familiar—Milwaukee, Portage, Wisconsin Dells. The train is commodious and comfortable. You can get up and walk until you get to a snack car, where you can buy almonds and cranberry juice. You can order an ale and play a card game in the club car. If you have the money, you can enjoy a fine meal in the dining car, at a table topped with a white cloth.

An Argentinian family sat down nearby and soon I was watching a laptop slideshow of their vacation in Patagonia while the fourteen-year-old son practiced his English on me with something like desperation. The first hundred pictures were great. When the slideshow ended I watched the countryside.

I disembarked in La Crosse during early evening and hailed a taxi, which charged twenty dollars to take me out to the closest car rental agency, at the airport. The rental car sped me southerly toward a hotel in Decorah, Iowa. On the Amish Byway near Harmony, Minnesota, I passed a family in a horse-drawn carriage, in which three children rode backward, looking out. Two of the children, one a girl, gleefully made the honking motion. I gave a light horn blow and was instantly rewarded with delight on their faces.

On my journey north I had done everything except hitchhike. Or ride a bicycle. Or a horse. The journey's conveyances consisted of feet, pickup, subway, bus, train, taxi, rental car, and feet again. Eight states. Thirty-six hours. I am thinking I should start a Slow Travel movement.

Only a special place would convince me to leave my own home and embark on a marathon journey. Heritage Farm, headquarters of the Seed Savers Exchange, was even more attractive than I imagined. My first impression was neat, straight, well-marked gardens in good-looking loam, with nary a weed. Volunteers at the visitors center were moving tables, and there I met another early arriver. I introduced myself and we strolled around, looking and chatting.

"Were you really chatting," my husband would ask, teasing me about a habit he finds charming but doesn't practice himself, "or were you finding out his life story?"

Honey, here's his story. The man is from Wisconsin and named Steven. He married late and well, and a few years into the marriage his wife was killed in a car accident only a few miles from home. Remember hearing that 50 percent of people get killed within five miles of where they live?

Steven sank into depression. That lasted a couple of years but he's getting better. He came to the conclusion that he had to move forward with his plans. He was glad to be at the seed gathering, although the trip was especially sad for him—he and his bride had planned to attend a seed-savers convention together, but never had.

"She would have loved all this," he said.

Around me, immaculate rows of leek, broccoli, lettuce, and beans were made even more endearing by the story of sadness and rejuvenation told to me as I walked along them. To the east a nine-foot hedge of Outhouse hollyhocks—a leafy screen polka-dotted with white, pink, magenta, and burgundy blossoms—bordered the garden. This classic is said to have marked the spot for the outhouse on Iowa farmsteads.

In these rows the Seed Savers Exchange devoted about six feet per variety, each marked with its name. Some of the plants I had never seen. One was a fantastical allium, with the blue-green tubes typical of the onion family, except at the tips of the spires grew what appeared to be another miniature cluster of bulbs, and from that cluster more stalks grew, and also, at times, at the top of that one grew even another cluster of bulbs. The plant looked like a four-story onion, rib-high.

The widowed seed-lover identified it as walking onion. Apparently the higher clusters get top-heavy and fall down, only to take root, grow into condominiums, fall again, and thus walk around the garden. I immediately wanted the plant.

By now more people had arrived, and my new friend and I were swept into a tour of isolation gardens. Somewhere along the way I parted from the group and ventured alone. I could cover more ground unencumbered.

The centerpiece of Heritage Farm is an enormous barn painted red with white trim. When I say enormous, I mean that a county fair would almost have enough room in this structure. A soccer game could be played in it. Its roof is high enough for a Ferris wheel! A second-floor door, through which hay was once tossed, hinges at the bottom because it must be lowered with ropes and pulleys.

Along one barn wall Grandpa Ott's morning glories crept up twine supports. This plant is an old friend. It was one of the first varieties I requested from a fellow seed saver, back in 1986. After a few years the plant was lost to me, as Latin speakers would say, meaning I lost it. I imagine the little garden I left behind at the edge of the north Florida woods in Sycamore. The spring after I left, some morning glories would have sprouted, most likely, since the morning glory self-seeds—but toadflax and Bahia grass would have begun to choke them and by the second spring, I imagine there was no bare soil for the few seeds to germinate.

That's why seeds need people. Domesticated plants, like many domesticated animals, can't survive without us. If enough people are growing a variety, then when one person drops the ball, others keep running.

Next to the incredible barn was a demonstration garden that, should you happen to see it, will spur you to yank out your monkeygrass. In small beds, leaf shapes contrasted with textures and colors. Even the spectrum of green was wide, from silverfish to sea to emerald to reddish. The effect was striking. In one small triangle, Dinosaur kale cradled parsley and verbena. In another, cucumbers twined around dill and calendula. A salsa bed blended tomato, cilantro, and onion. Three basils—holy, purple, and Thai—teamed up with artemesia. A Thomas Jefferson bed showcased plants Jefferson collected: hyacinth bean inching up a bamboo trellis, Tennis Ball lettuce, sensitive plant, Red Spider zinnia, Brussels sprouts. In other beds could be found interesting botanicals: Aunt Molly's ground cherry, strawberry spinach, and Norwegian Soup pea.

I stand guilty of gluttony in my love of plants. I am blameworthy of lust. Of jealousy too I am culpable. In gardens have I sinned.

I found my people at the Seed Savers Exchange. Becky Pastor of St. Louis, for one, stood behind an exhibit at the visitors center. She had begun a project called "Becky and the Beanstock," a blog covering the cooking of beans, with a recipe weekly.

"Most people can name five or six kinds of beans, and grocery stores carry about a dozen varieties," she said, "but there are about four hundred kinds of beans. The literature is mostly about how they grow. I wanted a lit that's as much about cuisine. I got together a group of friends who love wine," she said. "Wine tasters tend to have good taste buds. We had a bean tasting."

After speaking with Diane Ott Whealy, cofounder of the Seed Savers Exchange, Pastor received a gigantic box of beans in the mail. She has been showcasing them online, a bean a week. "Each has a distinct character," she said. "I think it's important for people to know what they taste like and what to do with them."

Her favorite dish?

"I really like succotash," she said.

I wandered into the barn to check out the seed swap. Only a few people this early had brought seeds to share. An elderly gentleman with a cane leaned on a counter where two tiny pepper plants, each about nine inches tall, rested.

"This is the grandfather of all chili peppers," the plantsman was saying. I drifted toward him. "Its seed was collected in Bolivia. I started the plants on March 31 at the Chicago Botanical Garden. *Capsicum chacoense.*"

The man was John Swenson. "Almost invariably in South American sites archaeologists find chile peppers," he said. "The three sisters (beans, corn, and pumpkins) keep body and soul together but there ain't much flavor there. So I call them 'The Three Sisters and Their Spicy Brother.'"

I wanted to ask Swenson some questions. I had a lot of questions. Top on my list concerned three piles of onions on a nearby counter, which bore a sign that said, TAKE SOME FOR FREE. The onions in each heap appeared to be related.

"Are these the walking onion?" I asked, nodding toward the bulblets, remembering my foray with Steven a few hours earlier.

"They are."

"All three?"

"Three different varieties."

"I've never seen it before today."

"They're called Egyptian onions, a corruption of *gypsy*," Swenson said. He was eager to teach. "If we don't know where something came from, gypsies brought it." He called the onion a garden cultigen, a cross between the common onion and perennial bunching onion. It's also called a walking, top-setting, or tree onion. He said there's no effort involved in growing it. In the old days, people peeled the bulbs for market. Sometimes farmers pull the whole plant and sell it as spring onions. "You will get many levels," he said. "I've had 'em four stories high."

"I'd like to take some," I said. "Which variety would you recommend?"

He pointed to a hillock of bulbs. "If you like to eat plywood, that's your plant." I won't name the variety Swenson dissed because there will be farmers who swear by it. Let's just say I chose a variety called McCullars White Top Set—one that, according to Swenson, tastes like a cookout on a summer evening.

While I was at the Seed Savers Exchange conference, I researched Conch cowpea in the one-room library tucked downstairs in the majestic barn. The latest (sixth edition) *Garden Seed Inventory*, which canonizes the commercial availability of garden cultivars, had a listing:

RUNNING CONCH COWPEA
90 DAYS
Non-clinging long vines, originally from which other cowpeas have been developed, harder to shell than modern varieties. Valued for its ability to resist insects and weeds. From the late 1800s.

The seed had been offered by one seed saver in 1991, one in 1994, one in 1998, and three in 2004. That had been only four years prior. So there was hope. I thumbed through a 2004 *Seed Savers Exchange Yearbook* and found that Running Conch had been offered by AL HA C, MO GE J, and PA WE W. I wrote down the addresses for Alabama's Charlotte Hagood, Missouri's Jeremiah Gettle, and Pennsylvania's William Woys Weaver, three more revolutionaries.

In the evening, after a tasty supper, Lynne Rossetto Kasper, host of the syndicated public radio show *The Splendid Table*, delivered the keynote speech. She is blond, with glasses, and wore a white blouse styled after a chef's coat.

"Microclimate is microculture," she said, and began to illustrate her point by talking about her Italian heritage. Real Parmigiano-Reggiano cheese comes from nowhere but the Emilia-Romagna region of Italy, where Parma ham is made, using only three ingredients—salt, air, and time. (Incidentally, if champagne is not made in Champagne, a region of France, it is not champagne.)

Kasper told the story of being in Bologna, the regional capital of Emilia-Romagna, where everybody ate *tortellini en brodo*, or pasta in capon broth. When she found herself in Parma, one hundred miles north, she ordered *tortellini en brodo*. "We don't eat *tortellini*," her host whispered to her, "It is a foreign food."

"Each place has a different collection of past history," said Kasper. "Where we live and what we live with is who we are."

In America we have a shortened history with food. Early Americans effectively displaced and decimated our native people, both co-opting their agricultural knowledge and snuffing out their agricultural evolution. The recorded, white story of food begins in the 1500s in pockets of this country. By contrast, the history of food on most continents reaches back thousands and thousands of years, with people staying more or less in place, practicing subsistence farming and developing unique cuisines.

Eating local in Italy, Kasper said, is more potent than eating local in the United States. Even the word *local* is different. For us Americans, *local* is geography: This came from a certain place. In Italian markets, some produce will be marked with the word *nostrono*, which is a possessive meaning "ours," as in "it belongs to us"; for Italians, *local* is more personal, a proud ownership.

Also in Italy, according to Kasper, exists a concept called *campanilisimo*, which is literally translated to mean "within the sound of the bell tower." In many Italian villages stands a bell tower, or *campanile*, and whomever lives nearby can hear it ring. Anyone not from within sight or sound of the *campanile* is a stranger. The Italian local, then, becomes even more focused and distinct, meaning anything produced within the sound of the bell that rings for you.

"Globalization is not doing any of this any favors," Kasper said. She got so worked up explaining what happened to food in America that she drew a big, jade-green fan from behind the podium and commenced to wave it. Somewhere we got the idea that food was science, she said. What worked in industry overtook what worked in the backyard, a kind of "better living through chemistry" mind-set. We let corporations make supper for us, not to mention breakfast and lunch.

Kasper's talk brings us to *terroir*, which has come to mean the relationship between soil or ground and the taste of a plant. This idea is based on the belief that the same plant grown in different places will taste differently. Take Vidalia onions. Vidalia is not a variety. A number of varieties of sweet onions are grown in a thirteen-county *territoire* in southern Georgia whose low-sulfur soil imparts a sweetness to the onion so incredible that people claim to eat them like apples. This is *terroir*, the taste of southern Georgia in an onion. *Sentir le terroir* is to smack of the soil, as in the native tang of wine.

I'm going to pause here, because I have another question for my friend Tom Stearns, president of High Mowing Organic Seeds. I want to know if the genetics of a variety change with different soils and environmental factors.

"Most definitely," he says. "Let's say you grow one hundred plants in Georgia. You save the seeds of the ten that do the best. I do the same up here in Vermont. For ten years both of us are always saving 10 percent of the best. At the end of ten years, if we plant your seeds side by side with my seeds, they will definitely be genetically different."

"Is this because of mutations?" I ask.

"They are not mutating," he says. "If you're growing a mustard green that can handle frost and you plant one hundred of them, some of them will go through the frost and survive. You've now gotten rid of a whole bunch of genes that make the mustard susceptible to frost."

"So this is what selection pressure means?"

"Yes, you're encouraging certain genes and discouraging others by the selections you make."

Kaspar's talk also reminded me of my amazing friend and inspiration Gary Nabhan, who put together a coalition of organizations dedicated to saving the remaining biological richness of our food system—not by locking it away in a vault, but by bringing it back to the table. He calls it RAFT, and has published a book with that title: *Renewing America's Food Traditions: Saving and Savoring the Continent's Most Endangered Foods*. What RAFT is doing, Nabhan said, is "trying to retain the synergies that happen when a particular plant or animal adapts to a particular landscape, soil, climate, and food traditions." He explains that traditional foods are a result of interactions between genetic stock and the soils and climate of a particular area. He calls them place-based foods. Working with RAFT, Slow Food USA began what it calls the "Ark of Taste," a catalog of over two hundred regional heritage foods that risk extinction because of industrial standardization. The goal of the ark is to bring endangered foods back to American tables, thus creating economic viabilities that will help them flourish again. Foods escorted onto the ark must taste great, be at risk biologically or culinarily, be sustainably produced in limited quantities, and be regional.

Seed savers are the raison d'être of terroir.

Okay, it's time to say that maybe the Seed Savers Exchange was not the mecca I thought it would be. There had been an upheaval in administration, a divorce between Kent and Diane Ott Whealy, and in October 2007 Kent had been removed from his job as executive director, a position he held since the organization began in 1975. In the ensuing years Kent Whealy wrote a series of angry missives to the Seed Savers Exchange board, sometimes sending copies to the entire membership, airing dirty laundry and accusing the gardeners' exchange of taking on a corporate mien.

Kent Whealy, who spent his career collecting a putative 26,000 unique varieties of heirloom garden crops, including 140 Native American varieties, accuses the organization of potentially giving away genetic material to corporations. When the nine-million-dollar Svalbard Global Seed Vault was finished in Norway in 2008, the exchange promptly delivered boxes and boxes of seeds. In a speech delivered in 2010 at the Land Institute, Kent Whealy said that depositing seeds in Svalbard places them "under control of the United Nations' Food and Agriculture Organization (FAO) Treaty, specifically designed to facilitate use by corporate breeders." He calls the participation a misappropriation. The *International Treaty on Plant Genetic Resources for Food and Agriculture*, which has not been ratified by the United States, states in Article 7, "The Depositor agrees to make available from their own stocks samples of accessions of the deposited plant genetic resources . . . to other natural or legal persons." Kent Whealy went on to say that "original samples," those stored back in the home country, are also covered by the order.

In a 2010 public letter of rebuttal, the Seed Savers Exchange board said that Kent Whealy's logic is "flawed at every step." The board admits that it is a "proud depositor" and "will continue to consign duplicate seed samples to the seed vault from our seed collection for safekeeping." In other responses, they assure their members that Seed Savers Exchange seeds are only in storage at Svalbard, belong solely to the Seed Savers Exchange, cannot be patented, can be retrieved by the board at any time from the vault, and will not be accessible by any other entity.

Why send seeds to Svalbard? Because it's a safety deposit box? Because it's backup? Because conditions are less than ideal at Heritage Farm? Because Iowa is subject to tornadoes? Because Norway is not as far away as we think?

I myself have been confused about the changing values of the exchange. In one example, the idea of a hand-to-hand exchange has deteriorated. Far more seeds are now sold to the public through the organization's full-color catalog than are exchanged among members. Once Seed Savers Exchange launched its catalog, members of the exchange saw requests, meaning the letters they got from people soliciting seeds, drop precipitously. Although the organization was founded on a gift society, meaning swapping of seeds among members, sales eclipsed barter. I was not the only person who witnessed this. Many seedfolk have reported similar declines. At least this means, however, as one old-timer told me, that the Seed Savers Exchange has developed a more stable system for keeping seed alive. If it's a corporate model, so be it. Maybe the time for the exchange is past, he said.

In a recent issue of a seasonal journal that comes with the purchase of membership, Seed Savers Exchange chides gardeners who "steal from the USDA plant germplasm labs." The USDA requires that seed requests from the country's official gene banks be limited to scientists and researchers. Some seed savers, the journal reported, have been obtaining germplasm through USDA gene banks by posing as researchers in their official requests to the USDA. Because these seed savers are gardeners and not scientists, the Seed Savers Exchange reprimanded them for writing fraudulent requests and thus for "stealing." The labs, they say, are not for gardeners who happen to have difficulty finding a certain seed.

That stance bodes ill. In my mind, the accessions in the USDA germplasm system belong to all Americans. They are *our* seeds, developed by our ancestors, grown by them and by us, and collected for use by our citizenry. Why wouldn't a populist seed exchange want its members to have access to public seeds? Wouldn't a gene bank *want* its seeds grown out?

The Seed Savers Exchange plays an important role in saving garden seeds, many of which were on the verge of extinction. But the controversy goes to show that as things grow big, they grow complicated—and often they grow out of hand. The controversy is a reminder, perhaps, of how important it is to constantly turn our attention back to the small, to the simple, to the local. And it is a reminder of the understated power, the unquestionable integrity to be found in a single, perfect seed.

Mecca or no, when the weekend ended, I made the long trip in reverse— the rental, the airport, the shuttle, the train station, Chicago, the bus, Atlanta, the subway, and finally the four-hour drive over familiar Georgia roads. Everything had gone perfectly. The hotel had had a vacancy. Its van had delivered me to the train station so I could return the rental early and save money. I had one clean shirt left. I turned off the road, which was still dirt, into my own yard. This was local—ours.

— 16 —

the pollinator

I AM STANDING in an Iowa field with Dave Cavagnaro, photographer and seed saver. He's a swarthy man with a Roman nose, thin as an insect. He's wearing jeans cut off at the knee, the corner of his wallet hanging out a hole in his back pocket, and a worn, plaid, short-sleeved shirt.

Six or seven gardeners hope to learn how to hand-pollinate squash from an expert, and an expert Dave Cavagnaro is. For eight years, he curated the seed collection for the Seed Savers Exchange. Now he is a botanical photographer and garden writer.

"I taught myself to hand-pollinate when I was eight years old," he says to the group. "It's easy."

Why would a person want to know how to do this? Because hand-pollination allows a gardener to grow many varieties of squash and at the same time maintain the purity of their seeds by eliminating (or at least reducing) the chances of cross-pollination. Hand-pollination allows the gardener to become a very discerning bumblebee.

Cavagnaro begins with the basics. "What we call squash are not just yellow squash and zucchini. By squash we mean the entire genus of cucurbits." (The cucurbit family includes many other plants, including watermelon, cucumbers, and luffa.)

He says squash are divided into four main species, plus two minor additions.

1. *Cucurbita pepo* has prickly stems and leaves, and its
   stems are five-sided: summers, crooknecks, scallops,

zucchinis, spaghettis, acorns, cocozelles, Delicata, vegetable marrows, small gourds, jack-o'-lanterns, and many pie pumpkins.

2. *Cucurbita maxima* has the longest vines and spongy, hairy stems: bananas, buttercups, Hubbards, turbans, Delicious, Hokkaido, marrows.

3. *Cucurbita moschata* is tropical, with large, hairy leaves and a widely flaring stem next to the fruit: butternuts, cheese types, sweet potatoes, Kentucky Field, Tahitian, tromboncinos.

4. *Cucurbita mixta* has slightly lighter leaves than *C. moscata* and the fruit's stem is less truncated—it includes most cushaws.

A fifth species, *Cucurbita ficifolia*, is a white green-striped pumpkin known as Malabar gourd or chilacayote from Oaxaca's mountains. Seeds of *Cucurbita foetidissima*, the sixth, known as buffalo gourd, are made into oil, but the fruit is not eaten.

Different varieties of squash within a species will cross-pollinate, meaning that a Hubbard will cross with a banana. That can be exciting if you're a plant experimenter or breeder. But in order to keep varieties pure and save their seed, a gardener either must choose to grow one squash from each species or hand-pollinate.

The sun was climbing up the forehead of the sky. "Well, we better get moving if we want to catch the flowers," Cavagnaro says. He looks around. "They definitely start to go limp at this hour of the day."

We acolytes trail him to the squash patch, where Cavagnaro teaches us to hand-pollinate. I had to watch him to learn how. But you don't have to. You can read and learn.

First, you plant some seeds of a squash you want to keep going. Then you watch the growing plant and note when it starts to put out flowers. Squash flowers are yellow. They are substantial, as flowers go—about four inches long and six inches in diameter, for the most part, at their zenith, about the size of a lily. There are two kinds of blossoms, female and male. The female flower has a small replica of the squash at one end. The male flower is plain.

The first bloom on a plant is usually not a female; squash put out a lot of male flowers before the first female opens. (Who knew that? Not me.)

Exceptions can be prolific bush summer squash, Cavagnaro says, which may make a few female blossoms before any males appear; those won't set fruit and can be picked with blossom intact and eaten young and tender. (Okay.) There's a male or female blossom at every leaf node. Once the plant starts flowering, this is usually the lineup on a stem: male, male, male, female, male, male, male. (Cavagnaro is obviously not just book-smart about plants. He's spent a lot of time paying attention in the garden.)

You go out and watch for the first female flower. You wait until it starts to turn yellow and looks as if it's going to open imminently. This is important. On the female flower, the stigma is not receptive until she opens. Recognizing an imminent blossom is not easy at first. Over time you'll figure out how a blossom looks the night before she opens.

On the night before the blossom should open, go out and tape the female flower shut, a chastity belt, and then tape shut a few male flowers, so that they can't open. To ensure greater genetic diversity, choose male flowers from different plants of the same variety. You will be hand-pollinating the next morning.

The next morning you go back out, untape the flowers, and pollinate by brushing the anthers onto the stigma. Go early. Squash blossoms are done by noon. Watermelons, which also attend family reunions with the cucurbits and are pollinated in the same manner, open later. The flowers of cukes, another member, last one day.

Cavagnaro leads us to a *C. maxima*. Sweet Meat squash is the variety, not a very exciting name. We crouch around the vines.

"Yesterday afternoon, around 5 p.m., I came out to the squash patch and located male and female blossoms," Cavagnaro said. "I sealed them shut with masking tape."

"I always open my males first," he continues. "Those bees can get in and out in a jiffy." He means that if you open the female, a bee can pollinate it with pollen from another variety of squash before you get the males opened. He picks a leaf like a surgeon and smooths it on the ground. Then he plucks three male flowers and lays them taped shut on the leaf. With each male flower, he carefully peels off the tape and tears off the hoop of petal. What's left are three small yellow rods an inch or so long in a row on the leaf.

The female flower Cavagnaro leaves on the vine, of course. He untapes it and parts its petals tenderly, until the stigma is visible. Then, very gently, he brushes each anther across and around the stigma. "Some people use a paintbrush to do this," he said. "That's the stupidest thing I ever heard. This is a gentle process."

He uses three males for the one female. "This ensures some diversity, if there is any," he says.

Once pollinated, he tapes closed the female flower all the way to its base. "Use masking tape, don't get painting tape," he says. "You need good, sticky, firm drafting tape." Of course the tape damages the delicate petals, but it doesn't matter because the flower will fall off when it sets fruit.

Last, Cavagnaro labels what he's done. He ties a length of marking tape around the plant stem—loosely, so as not to constrict. On the tape is written the squash's identification number and the date, then 3♂1♀, which is Greek for three males and one female blossom. The symbol reminds a young woman who is watching about something she's heard—that the sign for female, a cross topped with a circle, means "rooted in earth."

"If you don't have masking tape," Cavagnaro says, "you can use a male blossom to cover the pollinated female. Turn the female petals inward, then cover the whole thing with a sheet of male petal." He demonstrates the process. Turning the female flower's petals inward makes it look like an ice-cream cone rather than the bell of a trumpet. To create a limp sheet of petal to use as a roof, Cavagnaro picks a male blossom and tears off the calyx. Now it looks like a crinkly funnel. He slits the petal and opens it to make a small rectangle. He utilizes the opened petal to cover the female flower, a bit like a diaphragm. As the petal-wrap wilts in the sun, it will more tightly cling to the female flower, barring admission to insects.

"If a blossom fills with rain, it will be ruined," Cavagnaro says now. "Try to get the first blossom of a plant pollinated before a rain."

Cavagnaro moves to another set of taped blooms and begins the process anew. "There's not a remote chance that all of these we're pollinating today will set fruit. The first successful pollination on a plant will take. Then the others fall off."

In fact, the odds of success with pollinating by hand increase by pollinating the first female flower and removing any fruit that chances to set afterward on subsequent female flowers. In some melon varieties, only a small fraction of pollinations set fruit. Sometimes you can pollinate a dozen female flowers before one sets. Sometime it's much easier to isolate than hand-pollinate.

Then the pollinator waxes philosophic. "Squash plants are smarter than people," Cavagnaro says. "They only take on one project at a time. The plant says, 'I've got enough to do.' If it's a big squash, like a Hubbard, it may only make one fruit per vine. "The squash says, 'I'll make a helluva squash.' Or it finishes one squash and then says, 'Now I can make another.' That's the psychology of squash."

the bad genie
is out of the bottle

ONE DAY IN 1998, a letter carrier stuck an envelope in Percy and Louise Schmeiser's mailbox. The return address was Monsanto.

The Schmeisers were not expecting mail from Monsanto, the multinational responsible for 90 percent of the world's genetically modified seeds. They had never planted Monsanto seed, nor had any other dealings with the company.

Percy Schmeiser was a farmer, yes. Since 1947, he'd been growing canola on the Saskatchewan plains of Canada. But Percy Schmeiser was more than a canola farmer. He and his wife were seed developers, saving the best of their crop year to year, slowly breeding seed adapted to the Canadian plains and to the microclimate of their own thousand acres.

All of that ended with the surprise letter.

The letter stated that, following an investigation, Monsanto had good reason to believe that Schmeiser had planted Monsanto's patented seed, a GM canola, without a license on 250 acres of land and that this violated Monsanto's proprietary rights. To avoid legal action, Percy and his wife would have to pay the world's largest producer of GM crops for use of their product. At $115 per acre, Schmeiser owed Monsanto $27,850. There were three additional dictates: Monsanto had the right to take samples from Schmeiser's crop for the next three years in order to test for their canola; Schmeiser was forbidden from disclosing the terms and conditions of the agreement (if it could be called that) to any third party; and Monsanto, at its sole discretion, had the right to disclose the settlement terms to third parties.

Surely there had been a mistake. Schmeiser had never planted Monsanto canola. He saved and planted his own seed. But Monsanto had came along and tested some canola growing out by the road, and the canola contained their patented genes. The Schmeisers had been afflicted with something known as "genetic drift," the billowing of seed-matter by wind from neighboring farms onto their own.

Percy Schmeiser explained this. Monsanto didn't care. Its patented genetic material had been found on Schmeiser property and the Schmeisers should pay. Refusing to be coerced, Schmeiser said no, sorry, and the case went to trial. Meanwhile, Schmeiser turned around and sued Monsanto for $10 million for libel, trespass, and the contamination of his fields with Roundup-Ready canola. That lawsuit did not proceed through the courts.

In 2001, the Federal Court of Canada ruled in favor of Monsanto, determining that patent law supersedes the rights of farmers to save and grow seed and setting legal precedent. One of the key factors in the decision was that Schmeiser "knew or ought to have known" about the GM canola in his field. In other words, a corporation like Monsanto that cannot control how its gene is spread is not responsible for it—farmers are. If Monsanto plants were found on the Schmeiser's farm, then the Schmeisers were guilty of patent infringement. By this time, costs in damages and legal fees were in the hundreds of thousands. Percy Schmeiser, a hapless farmer, had fallen victim to the strategy of multinationals to gain control of our seed supply. Farmers in the United States have faced similar legal battles, one against Indiana soybean farmer Vernon Bowman, in which Monsanto won a similar ruling.

Schmeiser didn't back down. His appeal to the Federal Court of Appeal was heard in May of 2002 in Saskatchewan. The appellate court upheld the earlier ruling, so Schmeiser asked to put his case before the Supreme Court. In 2004, Canada's Supreme Court ruled that Monsanto's patent was valid but that Schmeiser did not have to pay penalties to Monsanto, because the farmer did not profit from the GM canola.

I heard Percy Schmeiser tell his story in 2005 when he traveled to Vermont at the invitation of State Representative David Zuckerman—a young Progressive Party member, organic farmer, and chair of the House Agriculture Committee.

Percy Schmeiser was a soft-spoken, seventy-four-year-old, unassuming man. He wore glasses and his hair was thinning. He stood on a low stage in the concrete-floored social hall of First Congregational Church in Brattleboro, before a card table that held his notes and a plastic cup of water. He spoke of the tremendous stress his family had endured, the debt they had incurred, and the breakdown of their rural social fabric as neighbor farmers who stood up for the Schmeisers received the same Monsanto letter.

"The whole issue for Monsanto is contamination," said Schmeiser. "It's like secondhand smoke." He adjusted the sweater he wore over a collared shirt. "Contaminate and people don't have a choice."

"The right of farmers to use seeds from year to year should never be taken away," he said, in a clipped Canadian brogue. "Some of the best wheat we have in Canada is developed by farmers, not companies."

The previous year, 2004, Schmeiser took a total of 161 flights in an effort to call attention to the hazards of GM seeds. "We're going to go down fighting for the rights of farmers," he said. "We don't want to leave a legacy of our land and our food full of toxins."

We can blame the wind. It steals pollen from Monsanto's flowers and brings it over into our fields. Then Monsanto comes after us, as if we were thieves, because it has found its patented genetic material among our crops. We didn't ask for the monster's babies to crawl into our arms.

Genetic drift is a handy lever to force farmers to use a corporation's seeds. And when it decides to whip a farmer into shape, into lining up to buy corporate seeds, it commences with threat. Then lawsuit. Out-of-court settlement. Or court.

By 2005, Monsanto had filed ninety lawsuits against U.S. farmers for patent infringement, meaning GM genes found in the fields of farmers that had not paid for the right, and Monsanto had been awarded over $15 million. I'll tell you here and now: We have a screwed-up justice system. These lawsuits and seeds are nothing less than corporate extortion of American farmers, said Andrew Kimbrell, director of the Center for Food Safety, as reported in the *Seed Savers Summer Edition 2005*.

Not only is the wind responsible for its invisible passenger, GM pollen, but so are the digestive tracts of birds and animals, our own clothing and shoes that attract seeds in crevices and hems, and the cheeks of mice and squirrels. All these spread genetic pollution. We can blame food aid too; somehow Oaxaca, hotspot for maize diversity, became contaminated with GM corn—despite Mexico's ban on it. By its very nature, pollen travels, and halting the advance of pollen from GM seed is impossible.

Imagine the scenario of transgenic contamination combined with the terminator gene. If you've never heard of a terminator gene, let me explain it. To prevent gardeners and farmers from growing out new patented varieties, scientists developed a method of rendering these seeds infertile, further busting up what has been for at least twelve thousand years a self-sustaining food supply.

Let's say you grow a zucchini. It's genetically engineered to taste like cotton candy. Every mother in the world rejoices because her kid is going to love this stuff. They go crazy for cotton-candy zucchini. So Mama Cuckoo, let's call her, saves the seed so she can plant a few hills of zucchini in the backyard. They don't come up. She plants more. Those don't germinate either! Little does she know, last year's plant contained a gene that causes the very seed she's trying to plant to abort itself, a terminator gene. She's trying to plant dead seeds. The only way Mama Cuckoo can serve more of the high-class zucchini her child raves about is a) to buy it from the grocer or b) to hurry to the hardware store and purchase some of the corporate seed. Either method involves a purchase. However, if Mama Cuckoo grew open-pollinated seed, she would be able to produce zucchini that tastes and looks and acts like zucchini. She could save their seed, plant the seed, and grow more zucchini—round and round, ring-around-the-rosy, without expenditure. Possibilities for her include Black Beauty, a variety found in seed catalogs from the 1930s; Constata Romanesca, a Roman ribbed variety; Mogango Liso, a round Brazilian one. She could grow Grey, Golden Bush, White Volunteer.

Use of a suicide gene begs a question: If hybrids don't necessarily grow true, why did companies need terminator genes? Again I turned to my pal Tom Stearns of High Mowing Organic Seeds for the answer.

"Many of the genetically engineered crops are not hybrids," he said. "They're open-pollinated. Soybeans, for example. There are millions of acres, billions of dollars. The terminator gene would prevent soybean farmers from saving their GM seeds."

At any rate, food activists decried the suicide gene. In 1999 the Rockefeller Foundation, which had funneled tens of millions into biotechnology research, asked Monsanto's board of directors to renounce it. In 1999, Monsanto chairman Robert Shapiro released a letter to the Rockefeller Foundation "making clear our commitment not to commercialize gene protection systems that render seed sterile." So Terminator was terminated, or at least we hope so. Geri Guidetti of The Ark Institute blogged in 2000 that over thirty terminator-type technology patents have been awarded and are owned by giant gene companies.

Schmeiser visited Vermont at a perfect time. The Farmer Protection Act, a state bill addressing farmers' property rights concerning seed, had just passed the Senate and had moved to the House. It had three tenets:

1. Seed companies wouldn't be able to sue Vermont farmers for genetic drift.
2. In the event a Vermont farmer was sued, the court case would take place in the state. (Monsanto had been forcing farmers to trek to their headquarters in St. Louis, Missouri, for court cases, at great expense.)
3. Most important, the person who owned the seed would bear the consequence of that seed. In Vermont at least, Monsanto would be responsible for genetic drift.

The Vermont legislature passed the Farmer Protection Act, but it was subsequently vetoed by the Republican governor at the time, Jim Douglas. However, many other states and municipalities also began taking local charge of GMO (genetically modified organism) contamination and also creating GM-free zones. In 2004, Mendicino County, California, inaugurated a ban on the propagation of GM crops and animals. Montville, Maine, banned GM cultivation in 2008. The movement gained so much traction that Monsanto began convincing states to pass preemptive legislation that prohibits such bans.

In 2005, the Schmeisers sent Monsanto a bill for $660 to sweep up more genetic contamination they had discovered. In 2008, Monsanto settled out of court and paid the costs of cleaning up the contaminated field.

The issue is bigger than the property rights of farmers. Genetic meddling is humans playing God, and GM technology was forced on America without us taking the time to fully understand its ramifications. For now, in this country, GM foods are among us, governed by companies that appear to be above the law.

Remember Percy Schmeiser. Put out of business by a little pollen floating on the wind, he lost almost fifty years of his farm's legacy. "There's no such thing as coexistence," I heard him say. "The GM gene is the dominant one. If you introduce GMs, there is no turning back."

We have not only introduced GMs, we have fallen heedlessly and haplessly into their traps. Are we going to be able to save the diversity of food? In this David versus Goliath battle, do the bad guys win?

tomato man

TOMATO MAN HAS A BIG GARDEN in a filing cabinet. He's got a little garden in the yard. He grows only tomatoes in his garden. He's got yellows and reds, oranges and streakeds. He's got stuffers and dryers, plums and cherries, currants and purples, grapes and pastes.

The garden in the filing cabinet is in little brown manila packets, tucked in file folders, everything labeled. It is tidy and well-organized, probably because Tomato Man is Dr. Charles Case, professor of sociology at Augusta State College, who has one foot behind a desk and the other behind a rototiller. In and out of the lecture hall, Dr. Case is a big believer in trimming the suckers. He even grades his tomatoes, and sometimes they flunk.

Although he should be expecting us, Dr. Case is not home when Raven and I arrive at a small, vinyl-sided house on one lot of land near Augusta, Georgia, 115 miles from our home. Beyond are first-growth woods. Soon Dr. Case pulls up. We're standing beside the real garden, looking. He says he's been to the store for Sevin dust, a chemical insecticide. His gardener, whom I'll call Jolene, is with him. This is her day to work.

"I bet you're amazed at how small this is," says Dr. Case. "When you wrote that you wanted to come see my 'operation,' I snickered. I don't have an operation." His blue T-shirt says AUGUSTA STATE STUDY ABROAD. On his face is a significant amount of gray stubble. He wears a Carhartt hat.

I read about Dr. Case in the *Seed Savers Exchange Yearbook*, where he was offering 312 varieties of open-pollinated heirloom tomatoes. The purpose of my visit is to see how he grows his tomatoes and get advice on which varieties to plant. The previous season I tested 22 open-source varieties of tomatoes in my hot, humid south Georgia garden. With the exception of cherry tomatoes, we did not harvest one fruit, mostly because of diseases—Southern blight and wilt—which are destroying our ability to grow our sweetheart vegetable.

"Since 1985, I have grown more than one thousand varieties of heirloom tomatoes," says Dr. Case. "Over three hundred of these have been highly successful. Well, let me show you around."

The operation is three different plots—one in the front yard, two in the back, the largest about thirty by sixty feet. One garden is shaded. Each tomato plant is from 2.5–5 feet tall, staked to a 5-foot-tall PVC pipe. The ground is completely weed-free, the bare dirt sandy and gray.

"I was pure organic," says Dr. Case. He must know by the way I look that I'm antichemical. "I tried various stuff. Texas Pete bug spray, soap. None of it worked worth a damn. So I've resorted to insecticides. We don't use herbicides. The weed control is rototilling and hand-weeding. And we use manure from the horses next door."

The plots are fenced to keep out deer, posts of PVC pipe strung with twine. Tied to the twine is a bunch of trash—large plastic bags, folded rectangles of tinfoil, and Styrofoam plates drawn with faces. Jolene—less than five feet tall, with plastic clogs on her feet—points out that each face is different.

"That's what you call a bored professor," she says. She is about fifty. Dr. Case looks to be in his late sixties.

"I know. I've got the summer off," he replies. He is trying hard to relax.

"So you have over 300 varieties of tomatoes here?" I ask.

"The secret to our 312 varieties is that we grow just a quarter of the collection, 80 varieties, each year. I've found that four-year-old seeds germinate about as well as new seeds."

"We grow four plants of each variety every fourth year," Dr. Case says. "Since they are self-pollinating, we're not so worried about gene pools." (He refers to the need to grow a minimum number of plants in order to maintain genetic diversity.) "Cross-pollination happens anyway from time to time. They claim bees will even force a blossom open to get at it."

Each plant is marked with a name; labels are plastic detergent bottles cut in squares, lettered with enamel paint that comes in a marker. One variety is already ripening. It is Red Alert. "This one is about fifty-five days," says Dr. Case. "For the first week that's your pride and joy." We walk along verdant rows, reading labels: GOLDEN QUEEN, BLACK MOUNTAIN PINK, COUSIN ROY'S STUFFING TOMATO, JITOMATE BULITO.

Either Jolene or Dr. Case has a story about each. Phyra has hundreds of cherry tomatoes. Napoli is a prolific paste. Liberty Bell is a stuffer. Reisentraube, "giant bunch of grapes" in German, is a drying tomato. Little Pink is a great tomato but it's yellow, not pink, although it blushes slightly.

The professor gives us a quick lesson on the two growth habits of tomatoes, determinate and indeterminate. "Determinate stems end with fruit and blooms," he says. "The fruits all ripen at once and then the plant dies. That's their strategy. They get done in a big hurry and die. At least that's the way they work for me here." He looks around and moves to a plant nearby. "Tip Top slicer, this is a prototypical determinate. And Wayahead here is perfect example of determinate too. Indeterminate stems end with shoots. They sprawl and tend to keep setting fruit until frost."

Not to be confusing, but it turns out that tomatoes are also delineated by the kind of leaf they have. Most tomatoes, like Eva Purple Ball and thousands of others, have the leaves we associate with tomatoes, flat and serrated. But others, like Prudens Purple, have foliage that closely resembles their cousins the potatoes—darker green, thicker, and somewhat puckered. Dr. Case tells me that potato-leafs have to be grown in isolation, because they will cross with regular-leaf tomatoes. They need one hundred feet between varieties to not cross. So he grows one potato-leaf variety per year. His darling is Brandywine.

"I've tried over one thousand varieties. I only keep the best," the professor is repeating. "It's my claim that I am offering the 312 best. If a tomato fails two or three years in a row, we don't try it again. This happened with Paul Robeson, for example. It would get to a certain point and die. We really wanted it to work." He says this because he's a race scholar; he wrote his dissertation on attitudes toward race equality.

It's hot out, very hot; in July Georgia is practically uninhabitable. Midway through the tour Dr. Case excuses himself and goes inside. Jolene is explaining that they start seeds of each variety in round pots, which they keep on the porch and move back to the dining room table when there's a chance of frost. These are planted two per two-inch pot. "We wait to transplant outside until we're positive there's no frost," she says.

Dr. Case comes back in a shirt that isn't sweaty; this one says AUGUSTA STATE BASEBALL. Jolene is telling us about neighbors who want to share in the tomato bounty. "We tell them we can't give away our tomatoes. We're growing them for seed. We started growing some hybrids the professor could give away."

"Tasteless things that made the neighbors happy," he says. "Let's go look at the seed collection."

Jolene has to wash glasses for us to have ice water. Dr. Case gets a twenty-four-ounce beer out of the fridge, then offers one to Raven, who's driving and politely declines. It's still morning.

The seed collection is in two filing cabinets. The first drawer is varieties with numbers for names. The second drawer says *ABC*. Dr. Case opens this one. File folders are lined front to back:

ABE HALL
ABE LINCOLN, ORIGINAL
ACE 55 STEAK TOMATO

Inside each hanging folder rests an envelope with notes on it and a few small manila coin envelopes (they cost 2.5 cents each, Dr. Case says) ready to mail to exchange members.

This is the dry garden and we start moving through it. Adventure, a red, says, "Impossible to exaggerate how good this one is." African, a pink-purple, says, "Keeps producing long after most have expired." Aviuri, tiger-striped red/yellow/green, "won prettiest tomato of the year at the office." Dixie Golden Giant is "huge orange-yellow beefsteaks borne in great profusion, mild sweet juicy refreshing taste."

Dr. Case is known for his descriptions.

The best description he ever wrote, he thinks, was for Bellow. It had that special taste, sort of funky. He wrote, "This is the kind of tomato that we grow heirlooms for." The next year he grew it again and it didn't live up to its description. Jolene is in the doorway. "It was good but it wasn't all that." Bellow's description now reads, "One of the strongest, most productive in my worst, most neglected garden."

"Maybe it had been a long time since I'd had a good tomato," he replied. He straightens from the cabinet. "Seed people are liars. If they say a cuke gets nine inches, it maybe gets to eight."

On each tomato envelope is a grade. Alteca 10 is B+. "I try to be generous but honest," says Dr. Case. Its description says "well-balanced taste."

Arkansas Traveler, an A+ ("world-class taste and texture") is deep pink. Some people think it's the same as Traveler, but it's not. They taste very different. Dr. Case pulls out a packet of seeds—"Here, put that in your pocket," he says to me. From the same drawer he gives me Black Mountain Pink ("great meaty slices with tiny seed cavities") and Big Italian Plum ("among the most flavorful of all plum types").

"How do you do taste tests?" Raven asks.

Jolene is the one who replies. "We stand at the kitchen counter. He cuts it. We smell it. We taste it. We drink water in between to cleanse the palate. If you get a really good smell, you know it's going to be good."

"We had one, Giraffe, bred in the 1990s at Russia's Timiryazev Agricultural Academy, that tasted nasty. It was huge in the garden, ten feet tall. But inedible."

In the other drawers are more of the same, rows of file folders filled with seeds of varieties—each with a grade, each with the best description that Dr. Case can conjure. He keeps his original seed packets sorted by year in the bottom file drawer. If his plants come out wrong, untrue to type, he finds the original seeds and starts over. I keep asking about favorites—I want to know what to plant, what's resistant to Southern blight—and they keep having a hard time with the answer.

"Well," Jolene asks, "am I going to stuff it, slice it, make sauce, or eat it in the garden?"

Finally they say Black Cherry is their all-time crowning-jewel of a cherry tomato. Its description says "hard to gather seeds because everyone keeps eating them." Dr. Case and Jolene also dote on Miss Dorothy ("super-rich classic old-time taste, keeps producing for months"). The story behind this variety, they tell me, is that Miss Dorothy Beiswenger of Minnesota sent a variety in 2003 labeled "special miscellaneous" with instructions to grow, name, and offer it. Dr. Case liked it and named it after "this wonderful tomato person." "MN BE D is who she is. She's an old lady now, you can tell by her handwriting." Jolene retrieves a packet of the seeds from the file cabinet.

"You need these," she says.

We go sit down at the dining table, finding room between ashtrays, papers, and books—a real mess. Maybe nobody thinks it's a mess but me. This is getting kookier by the minute and I love it.

Dr. Case pulls out a spiral notebook full of lists written by hand on legal paper. Dr. Case is a prodigious list maker. These are long lists of varieties and the date they were planted in the ground. From 2002, there is a list of 149 varieties. From 2003, there are 156. In 2006, 150. He notes that his handwriting is getting better over the years. This year's plants, I see, were put in the ground on April 6.

One of his lists is how many varieties of tomatoes people offer via the Seed Savers Exchange. He's been in the top ten for a few years. Neal Lockhart, IL LO N, has over 700 varieties. Bill Minkey, WI MI B, has 661. "I'm number 5," he thinks. But we find his list for the previous year and he's only number 7. TN JO M (Marianne Jones) is number 4 with 565. IA DR G (Glenn and Linda Drowns) is number 5 with 417.

"I have a bit of competition in me." Dr. Case laughs. He holds his Beck's Light by the neck and chugs it. These aren't sips. These are long pulls. When he tips the bottle down, one-fifth of it is gone.

Thumbing though his legal pad, I see a "Best of 2007" list. This may be what I'm looking for, so I scribble in a hurry:

Arkansas Traveler ("brilliant rose pink/red inside and out")

German Giant ("deserves lots more attention")

Olympic Pink ("four-foot vines that grow with the determination of a bulldog")

Purple Brandy ("at least twenty pounds per plant, but the taste is the thing, deep rich funky old-time flavor"—this one is a cross between Brandywine and Marizol Purple, bred by Joe Bratka)

Eva Purple Ball ("if you try one, this should be it")

Jolene gets up and goes out. It's time for her to leave. Dr. Case says something to her that I don't hear, something about Big Mama beans, and she says, "Okay, Professor."

Dr. Case wants to start a small seed company one day, selling these heirlooms. He's got the math figured out. The average tomato has 250 seeds, although some tomatoes, particularly paste, are famous for being stingy. A plant produces between 4 and 30 tomatoes, an average of 10. At 10 tomatoes, each plant produces 2,500 seeds. Each packet Dr. Case sells contains 25 seeds. At $2.50 that's 10 cents a seed. At that rate, he figures, he has around $80,000 worth of seeds in his garden.

"Gosh," I say. "How many do you sell a year?"

"Oh, shit," he admits. "We sell about $500 a year."

I want to hear the math again. In a given year he has about 320 plants, 4 each of 80 varieties. At 10 tomatoes per plant that's 3,200 tomatoes. Given 250 seeds per tomato, Dr. Case would harvest 800,000 seeds. And at 10 cents per seed that equals $80,000 worth of tomato seeds.

But the money is secondary, he says. Preserving seed is the first mission. They squeeze out the seeds, then make sauce. Every year they freeze twenty gallons of sauce, which they give away. Every batch is different.

Jolene's ride arrives. She pops back inside and kindly interrupts to ask the professor if he will pay her. It will be fifty dollars, she says. He pays her, she says goodbye, and she kisses him casually on the lips.

After Jolene leaves he tells us she has tattoos all over. She was a biker chick. Her husband, maybe it's ex-husband, is in prison for killing two men, a drug deal gone bad. She has two daughters, Brandy and Sherry. He met her while she was bartending; he was her customer.

"You've got a good partner there," I say.

"Definitely."

I steer the conversation back to seeds. With seven thousand varieties of tomatoes in the world, how did he choose?

"This borders on strange," he admits. "I started planting only five-letter varieties. I tried 120 of them and about 50 worked."

"Why five letters?"

"Well, here's the logic of it. First of all, descriptions are exaggerated. So random selection is as good as any. Three letters or four are too few. Five offers economy in writing—you wouldn't believe how many times you have to write the name—nametags, labels, notes, and so on. What about a name like Norinka Pridnestroviya? Heidi is much simpler. One of my favorite five-letter varieties is Peace."

"You might forget that I'm an intellectual," he continued. "I've already expressed my cynicism toward exaggerations. So this is essentially a random sample. I could have simply picked a letter, like M. But I chose length of name, initially. I have 100–150 five-letter varieties."

I am thumbing through *Seed Savers Exchange Yearbooks*. Dr. Case offers Amaze ("inside is radiant rose pink"), Tiger ("a personal favorite"), Black ("dark mahogany brown with green mottling"), Wihub ("plum-shaped fruit"), Venus ("an image of loveliness"), Omara ("multipurpose"), Mayan ("a pot of sauce on each plant"), Fakel ("try it despite the name"), Dusky ("stunning beauty"). I'm sure I could find at least one hundred varieties with five-letter names.

"But I abandoned that system," he says. "Now I just pick varieties whose name or description appeals to me."

It's almost one o'clock and Dr. Case's second beer is done and I can't think of any more questions. I thank him for taking time with us, for answering all our questions, and for giving me some nice seed. I tell him that he's been a pleasure to talk with, and I'm impressed by his operation and his devotion to heirlooms.

"I'm just a crummy sociologist," says Tomato Man.

how to save tomato seeds

PICK NICE TOMATOES that would be perfect for a mean kid to mash up. If they're large, slice them in half at the equator. Hold them over a canning jar. (Try not to use plastic for anything. Plastic is bad stuff.) Milk the pulp, meaning the gelatinous matrix that suspends the seeds, like frog eggs, into the jar. If you're working with cherry tomatoes, you'll have to hold the whole tomato between your fingers and squeeze. The only thing left will be the skin.

Put the jar lid on, give it a shake, and label it with the name of the variety inside. If you don't label the jar, you will forget what it contains. If you have two tomatoes you're saving, you think you can sit Yellow Mortgage Lifter on the right and Pruden's Purple on the left and remember what's what, and pretty soon you're wondering if Yellow Mortgage Lifter was on the right or the left. Just do it.

The tomato hull can still be eaten. I think sauce is a good idea at this point.

Fermenting, which is what you are doing with the goopy mess in the canning jar, is the best way to save tomato seeds because the process dissolves the gel—which contains chemicals that inhibit germination. Fermentation causes the seeds to germinate more quickly when you plant them the following spring. Fermenting also breaks down the seed coat where seed-borne diseases like bacterial canker, spot, and speck can lurk. Let the mess stand for two or three days in a warm location, longer if the temperature is below 70°F. The books say to stir daily but I don't.

When a layer of blue-gray mold covers the surface of the tomato-seed funk, the process is complete.

Occasionally in hot weather (seven months a year here), I have had the seeds start to germinate inside the goop, which means that I've left them too long untended and they think they've actually been planted and it's time to race off again into plant-building and fruit-making. Don't be like me.

Look at the underside of the jar. The viable seeds will have sunk to the bottom. Pick off the scum, then fill the jar with warm water and begin to pour off the now-rotten goop, being careful not to pour out your seeds. You may have to add water or rinse seeds off the insides of the jar and pour again, slowly. Viable seeds keep sinking to the bottom. Do this until you have mostly seeds and water in the jar.

Now dump the seeds into a large metal strainer whose holes are smaller than the seeds, rinse, drain for a few minutes, then spread them on a screen or on a plate covered with newsprint or a clean rag (don't buy paper towels). Leave the seeds until they dry.

Label—very important!—and store.

— 20 —

sweet potato queen

WHEN I THINK SWEET POTATOES, I think of Yanna Fishman. And I think of Yanna often.

Yanna is the kind of woman anybody wants to meet when they are dabbling in a passion and need to talk with someone who really knows the subject. That's the way it was between us. Like the blind dog who sometimes finds the bone, I made acquaintance with the expert.

Yanna's husband is Doug Elliott, a folk troubadour and author I met at a land and water conference in Pennsylvania, where we both appeared on the schedule. He told me that his wife was a seed saver and that her specialty was sweet potatoes.

In reality, keeping sweet potatoes from year to year does not require the saving of seeds. The tuber serves as the seed. From year to year a gardener must store a few sweet potatoes. When spring arrives, the potatoes begin to sprout and the sprouts, called "slips," are planted.

Up to the point I met Yanna, I'd been buying sweet potato slips from the local hardware store. I suspected, however, that sweet potatoes would be similar to any other American foodstuff, that vintage varieties would need stewardship. I wrote to ask her if she could recommend a variety for me and launched what turned into a battery of questions. I thought I'd made a simple request, but I soon realized just how little I knew about sweet potatoes.

"What color do you want?" she asked.

"What colors are there?" I thought all sweet potatoes were a deep orange.

"Well, lots. Red, yellow, white, gold, purple."

"There are purple sweet potatoes?"

"Yes," she said delightedly.

"Whatever color tastes good," I said.

"What kind of taste do you like?"

"Sweet," I said.

"How sweet?"

"Different potatoes have different sweetnesses?"

"Very much so."

"The sweeter the better."

"And what about texture?"

I'd never thought about texture and I said so.

"Some are more dry, some watery, some bread-like, some creamy."

"Creamy and moist," I said.

"And do you need a good storing potato?"

"I guess so," I said. I was far out of my element. But I wanted to learn more about this amazing plant scholar.

In the heat of July 2009, I visited Yanna in her wild garden. I was surprised to find a short woman, up to my ribs, with long graying hair and sweet brown eyes, wearing jeans and a bluish plaid shirt. Yanna lives in one of the regions of highest agrodiversity in the country, the highlands of western North Carolina. When she first moved there more than two decades ago, she was interested in world-change gardening and began to listen at community suppers for talk of heirlooms and to ask her neighbors about their history with varieties. Sweet potatoes grow especially well in western North Carolina, and Yanna soon learned that the crop had been vitally important in the economy of the region. She learned to ask two questions of the farmers around her: What potato do you grow to sell? What potato do you grow to eat? She began to collect both the germplasm around her and to order unique varieties from gardeners around the county.

"About 80 percent of sweet potatoes grown commercially," she said, "are common varieties—Beauregard, O'Henry, Porto Rico. But there are hundreds and hundreds of heirlooms."

In two medium-sized plots in the Carolina hills, Yanna grows more than forty varieties of sweet potato. Sometimes she invites friends over for potato tastings. She picks twenty varieties, punctures the letters of the variety name in the potatoes themselves, and bakes them. When she slices and serves them, she labels the plates with the varietal names. The judging criteria is by ranking from 1–5 for sweetness and texture, and guests are asked to describe the taste ("watery," "dry," "starchy," "chestnut-flavored.") Yanna is the Sweet Potato Queen.

Every story has a substory, sometimes many substories, and one of the substories of Yanna is generosity. In my investigations, generosity is a trait that I've found almost ubiquitous in seed savers, many of whom realize that in order to preserve genetic diversity, seed must be shared. They also seem to realize that we need people to become passionate first about gardening and then about sharing seed, and that sometimes a gift sparks a passion. In fact, maybe generosity is the story and seeds are the substory.

With Yanna, generosity is built into almost everything she does. At the potato tastings, for example, she makes notes of friends' preferences and, come spring, bestows a handful of slips on them. She had neighbors who had grown an old variety named Nancy Hall, but had lost it. "I got it back for them," Yanna said. "It's not the best sweet potato, but they're happy growing the variety their parents grew."

"What do I owe you for these?" a neighbor asked Yanna after one such gift.

Yanna replied, "What do I owe you for all the things you've taught me?"

With Yanna, I wonder if her choice of sweet potato as a vegetable to embrace isn't significant. The sweet potato comes close to being a perfect food crop: long-storing, nutrient-packed, easy to grow—and most of all, sweet. It's a generous plant.

"Why sweet potatoes?" I ask her.

"My son can't get enough," she said. "To keep him in food, I bake him a panful. That's his staple." It seems too easy an explanation, but Yanna has moved on. She emphasizes that she grows only what her family likes to eat. "I'm a gardener slash cook. I cook what I grow and I grow what we eat. That's why I don't grow okra and grow only two squash plants." At first, she said, out of responsibility to the gene pool, she grew twenty-five plants of every sweet potato variety she collected. "Now I grow way more of what produces well and tastes good and plant five to ten of all the others."

For over a decade Yanna has been corresponding with Ken Pecota, North Carolina's sweet potato breeder. Sometimes he sends her slips of new varieties—"Right now he is working on purples for anthocyanins."

In turn, Yanna sends Pecota her garden records (she started keeping them in 1988)—"We compare his slips to my slips." For many years she noted output but didn't count the slips. "I had a blinding flash of the obvious," she said. "What does output tell you if you don't count the slips?"

Yanna's garden is asprawl with vines and when I look closely, I see that not all sweet potato leaves look the same. Some are lobed, some almost ferny, others are entire. Each sweet potato variety consumes a few feet of space. Hernandez, a juicy sweet potato that Yanna says is grown for the baby food industry, is growing next to Hayman.

In 2010 Yanna nominated the variety Nancy Hall to the Ark of Taste. "While not the most productive of my varieties," she wrote, "it has a rich golden color, firm texture, and delicious flavor." In a Texas Agricultural Experiment Station publication, she found a 1895 reference to the Nancy Hall. Although like many vintage vegetables the origin of the potato is unclear, an 1895 letter written by A. J. Aldrich of Orlando, Florida, claims that the variety came from an accidental planting by a Miss Nancy Hall. The seed were mixed into a packet of seeds. By the 1930s and 1940s, it was one of the most popular varieties in the South. Now Miss Nancy Hall has boarded the ark.

Yanna Fishman doesn't dally with seeds. She's not a piddler. To understand how much effort her sweet potato project takes, I must tell you what is involved. In the fall, Yanna harvests the forty-plus varieties of potatoes. She divides each variety into two buckets, the smallest ones from the best plants for seed stock and the rest for eating. In actuality, she uses three buckets; the third is for travelers—potatoes grown on vines that spread beyond their bed and whose origin is difficult to determine. While harvesting she notes the variety with the best hill of potatoes, the variety damaged least by insects, the one most heat-tolerant, and so forth. One constant running through all her efforts is her meticulous record keeping. A stick with the name of the variety printed on it in permanent marker goes into each bucket. The potatoes are then spread on bread trays and crates all over the yard for a couple of hours, where they dry and are cleaned off.

The sweet potatoes are transferred to paper bags—eating bags and seed bags—labeled with the year and the variety. Yanna weighs the bags to see how productive the vines have been. "Some get three pounds per slip," she says. "I average one-half to one pound per slip."

"How many potatoes do you save for stock?"

"It depends on the variety," she says. "Some make more slips than others. Often I save all that look good."

The potatoes cure in the bags in her greenhouse. "They like it hot and humid," Yanna says, "90 degrees, 90 percent humidity. There used to be sweet potato curing houses up here where people would take their sweet potatoes to get the curing done quickly. The owner of the curing house would take a portion of the potatoes. They were a big crop."

After the curing, she stores the bags of potatoes in her pantry. On the spring equinox or thereabouts, preparing for spring planting, Yanna beds the potatoes. She makes a trip to the local sawmill for sawdust, then fills an assorted collection of buckets with the moistened shavings. The tubs stay in the greenhouse until no longer threatened by frost. "This begins the Sweet Potato Cactus Wars," said Yanna. "Cacti that have been living luxuriantly in the greenhouse start getting moved all over the house."

When the potatoes sprout, sending up green shoots from the tuber itself, and when Yanna is ready to plant, she cuts the slips two inches above the potato and holds them in yogurt containers labeled with the name and year and filled with water. These go into the ground and are liberally watered for a few days. Then for the growing season she mulches and weeds and waters and cares for the plants, until it's time to harvest and the process begins all over again. That's the life-cycle of a seed potato.

Besides over forty varieties of sweet potatoes, Yanna grows other endangered vegetables—Hercules cowpea, Purple Knucklehull cowpea, Greasy bean, and a spice pepper that came home from St. Croix with her lunch. She plants herbs to make tonics; one friend credits the teas for her long-awaited pregnancy, Yanna told me happily.

Throughout the orderly beds volunteers spring up helter-skelter. One is Hopi amaranth, which will turn bouquet water a lovely purple. Another is Bright Lights cosmos, like flaming stars or miniature suns in the garden, which Yanna uses to dye wool that she spins and knits into hats. Bright Lights has come to live at my house. There is magenta lamb's-quarter that came from the neighbors. There is shiso, with which a Japanese visitor showed Yanna how to wrap sushi.

Everything has a story. Yanna says she lost the tithonia (Mexican sunflower) once. For a couple of years no volunteers sprouted and she had not saved seeds. But she had given the tithonia to her neighbors, the Websters, and she was able to get it back from them. Yanna's son, Todd, has joined us. He is a healthy and bright teenager with a lot of energy, eager to interact. "A similar thing happened with evening primrose," he says to me with a sparkling grin. "Jimmy Cooley gave it to Russell Cutts who gave it to Denise McClellan. The gift keeps moving but not in the direction it came." Todd calls the reciprocity "giving it forward."

Despite her boundless generosity, Yanna does not offer her seeds through the Seed Savers Exchange, although she is a member. She only shares interpersonally. "Everybody who comes, if they like a plant, I give them seeds," she said. "I like to follow through." Yanna sometimes attends the seed swap of Southern Seed Legacy and one year was named its seed saver of the year.

Luckily Yanna invited me to spend the night. I got to see her jars of seeds, her bags of potatoes, a rotational chart for tomatoes handwritten in pencil. We talked into the evening, before I was shown to an artful guest room above the barn.

The next morning Doug cooked breakfast outside on an open fire. The morning was beautiful, cool and softened by dew and wispy fog. Soon we were seated at a picnic table between a greenhouse and a grape arbor, eating steaming platefuls of venison, scrambled eggs with broccoli, bread with honey, and potatoes. Everything was homegrown or local.

By the next spring, I had determined one of my own criteria for a good sweet potato—one that is creamy, one that is not too starchy, and one that when baked exudes a sugary syrup, which we called tar when I was young, across the baking pan. That seems to indicate a really sweet potato. I remember the sweet potatoes I knew as a child being full of tar and this trait getting harder to find in modern varieties.

As for all the other possible traits I asked for, Yanna chose two varieties she thought I'd like. Soon an overnight package arrived with a bundle of damp slips of Ginseng Red and another of Red Gold. Yanna included a third bundle of slips, a medley of many different varieties that I could plant to see if I especially liked something. The slips grew superbly and filled four garden beds, then overran the beds and traveled through the garden. At the end of summer, I grabbled a couple of Ginseng Red without pulling up the hill and baked them.

They ran tar across the cast iron skillet. They were deep red, creamy, and so sweet they'd make robins sing.

I remember the late fall day that I harvested all the potatoes. The sun was soft and golden in the sky, its rays angling from the southwestern horizon. I began to dig the medley. I was using a pitchfork, wielding it carefully so as not to damage the tubers. The first sweet potato I unearthed was smallish, six inches long. I brushed off a layer of soil and found the potato orange-skinned, flashy as a carrot. The next one was bigger and white, really a creamy shade of ivory. Then there was a slim purple one. On and on, each potato was different from the one before until I had a box of sweet potato crayons, an array of earth-colored armadillos, lumps of beautiful clays. After a while I bundled up the spent vines and dumped them into the goat pen. I carried the next bundle to the hogs.

Afternoons like that I hope I never forget.

keener corn

I HAD HEARD that a man near Rabun Gap, Georgia, was growing an old corn. The man's name was Bill Keener and his address said Betty's Creek Road, with no number. One fall day I found myself up in Rabun Gap, in the foothills of the Appalachians, with an unscheduled morning, so I went looking for Betty's Creek Road and headed up it. I came upon a garage where a receptionist was working who knew Bill Keener and told me where he could be found.

The high diversity of heirlooms in the Appalachian region was established by Jim Veteto in his doctoral research at the University of Georgia. He credits the Eastern Band of Cherokee as the originators of much of the diversity in the Mountain South.

When I pulled into his yard, Mr. Keener was washing a small truck.

"Need some help with that?" I, a stranger, called as I got out.

"I'm about to get it," he said.

"I'm looking for a man named Bill Keener," I said.

"You've found him."

Mr. Keener seemed happy to stop work and talk to me about old-time crops.

"Just a moment," I said. "Let me get paper to write on."

He turned off the spigot. "I don't want to be in the news."

"What about a book?"

"Maybe that's okay," he said, so I went easy, lobbing him soft questions, the weather first and if he made the birdhouses nailed on trees all around.

"Yes."

"Is that tree a pear?"

"It is."

Mr. Keener and I settled into a pair of lawn chairs in the yard. "I was sent by Woody Malot and Cary Albright," I said. "I'm looking for old varieties of vegetables. They said you have an old corn and I'm interested in hearing more about it."

"What do you want to know?"

"Is that it growing yonder?"

"That's it."

"That's a tall corn. Ten or twelve feet tall."

"And only one ear per stalk."

Even before I drove up Betty's Creek Road that day, I had entertained ideas of collecting this corn, since I knew Mr. Keener was aging and somebody needs to keep his family heirloom alive. When I heard that the corn only produces one ear per twelve-foot stalk, however, I immediately lost interest. If it was left up to me, this variety would go extinct, I guessed. Part of me wants to save everything. Another part of me wondered what good a corn can be that only bears one ear per stalk.

"One ear?" I exclaimed.

"It used to have more," he said.

"What happened?"

"I don't know."

"Has it always been that tall?"

"It has gotten taller."

"Where did you get it?"

"Aw, my daddy grew it. His daddy before him. Maybe his daddy. I've grown it all my life."

"Does it have a name?"

"We knew it as Keener corn."

"Do you mind showing it to me?"

"Not at all."

A tall man, Mr. Keener plucked himself from his lawn chair and sauntered across the mown grass to his quarter-acre garden that had been plowed under except for a few rows of corn and some of beans. The corn had matured and dried on the stalks.

"Why haven't you harvested it?"

"I never pick it until November. I want it to be good and dry."

In the corn patch, we walked among giants, high above us the tassels like tan fingers of tall skeletons.

"I'll give you a couple ears," Mr. Keener said.

"You don't have to," I said, dubious. "Keep your corn for grinding. You don't have that much."

"Oh, I have plenty. I have another garden full."

"Well, I'll take an ear. But just one."

"Oh, you'll want it," he said. "No corn makes better meal than this." Mr. Keener approached a stalk. I saw no ears on it until I looked up and saw one above my head. This corn didn't need scarecrows. It was so foreboding that *it* would scare the crows away. Mr. Keener reached up and tried to break off the biggest ear of corn I'd ever seen. From dry silks to the stem where it attached to the stalk, it was nineteen or twenty inches long. It was like a club. And it wasn't detaching easily. Mr. Keener bore down and wrenched on it. He handed the ear to me like he was handing me a baby and reached for another ear.

"One's enough," I said to him. I felt guilty taking corn that grew one ear per stalk. Secretly I hoped he'd give me three or four ears, enough for diversity, because I'd changed my mind, that quick.

"If you're going to grow it, you need at least two ears," he said, as if reading my mind.

"You are very kind." That day four corn plants sacrificed their entire year's labor to me and I was thrilled.

"Every time I hear that the neighbors are going to plant corn, I get nervous," Mr. Keener said.

I was intrigued that Mr. Keener brought this up. "Why?" I asked.

"Because it will contaminate my corn," he said. "I'm afraid of the GM corn getting in here. And most people don't plant anything but that."

This man was almost eighty years old and he understood genetic engineering.

"Once I planted one of the modern corns right alongside Keener corn," he said.

"What happened?"

"It took me almost ten years to back it out."

"How'd you get it out?"

"Selection," he said. "The Pioneer corn had a different look."

We stashed my vintage maize in the car and I knew that I had to do something quickly or my visit with this interesting and outspoken man would be squealing to a halt.

"I'd love to see your setup for grinding," I said.

"Let's go take a look."

In a small and dusty barn, Mr. Keener showed me two corn shellers, one a hand rig and the other electric. The floor of the barn was littered with long cobs. Some of the cobs were red and some were white.

"What's this about?"

"Years ago I introduced a Tennessee Red into the corn," he said.

"Was it GM?"

"No. It was open-pollinated too."

"Why did you do that?"

"I thought it needed it." That's all he would say. Really, what difference would it make? The Keeners have been keeping this corn for generations. He can breed his corn however he likes and it's still Keener corn.

"You don't see much evidence of the Tennessee Red except for a few red cobs," he said. "This corn makes fine cornbread. I've got people coming back for it year after year."

"Because they like the taste?"

"You will too. And because I nub it."

"Nub it?"

"I take off those hard little kernels on the pointy end of the cob before I grind it. My daddy showed me how to do that. But I invented a better way."

"May I see how you do it?"

He bent and picked up a metal cylinder a bit smaller than a full ear of dry corn. "That shouldn't be on the floor," he said. Inside the cylinder were welded metal wedges with little teeth cut in them. The wedges followed the conical shape of the tip of a corn ear. Their teeth take off the tip kernels, which should have a name if they don't already. "I call it a nubber," he said.

"You made this?" I asked.

"I did."

"That's genius."

I'll admit here and now that I coveted everything the man showed me: the corn patch, the barn, the corn sheller, the crib he had lidded with mouse-proof hardware cloth, the nubber. I coveted even the cobs strewn on the floor.

Before I left, Mr. Keener picked a mess of Greasy Back beans, an old-fashioned Appalachian cultigen so named because their pods have a greasy appearance, good for snaps or shellies, which were drying on the vines. As well, Mr. Keener gave me a few half-rotten tomatoes of a heirloom variety he called Box Car Willie, an orange-red beefsteak with average yields.

I left with three new pets, yes.

getting the conch back

WHEN I RETURNED HOME from the Seed Savers Exchange convention I called Jeremiah Gettle at Baker Creek Heirloom Seeds, one of the people who might still have Running Conch cowpea. Gettle was busy, so I spoke with the seed curator. "I'm looking for Running Conch cowpea," I said. "I don't see it in the catalog."

"Definitely we used to grow that," the curator said. "Let me look for some seeds and see what I can find." A few days later he called with bad news.

"I don't know what happened," he said. "We have none of these. This is unusual for us. I've made a note to find them and start growing them again."

My heart sank. I had two more chances at redemption.

I found the phone number of Charlotte Hagood in Albertsville, Alabama. "I know you haven't listed Running Conch for a few years," I said. Nobody has. "I'm eager to get my hands on some seeds. I'm calling to see if you have some stored."

"I'm sure I do," she said. "The seeds are in the freezer. Next time I go in, I'll pull out the Running Conch."

"I'll go ahead and send you a formal request," I said. The Seed Savers Exchange requires that a few dollars accompany each seed order, along with a self-addressed stamped envelope. "Then everything will be ready when you find the seeds."

"That's fine," she said. "It may take me a month or two. But you won't need them for awhile." She referred to spring still being months away.

"I very much appreciate it."

By January the seeds had not arrived and I redialed Charlotte. She was busy—a characteristic, you'll notice, of revolutionaries—but said she hadn't forgotten the seeds and would be sending them before much more time passed.

What she didn't know about was my desperation. She didn't know that I had grown the seeds and lost them. She didn't know that I was afraid she would check her supply and find out that she'd been mistaken, that she had lost them too. I was afraid that they were gone for good, a big X-mark on my karma.

To cover myself, I phoned William Woys Weaver. Definitely he had them, he said. Send an order and he'd fill it. I did and he did, and about the same time that Weaver's cowpeas arrived, a packet from Charlotte also materialized. I was pretty excited to get the cowpeas. Right away I untaped a package to see again, after so long, the seed that had fueled my concern. The peas were as tiny as I remembered, opalescent, like the moons of Jupiter. They were like long-lost cousins. They were like hundred-dollar bills and I felt as rich as I've ever felt.

"Twice the genetic diversity," I thought. Using the same variety from different sources, at least initially, could only strengthen the strain.

I wasn't saved yet. I had to grow Running Conch and save it. I had to do it consistently, year after year. I had to make a commitment and be faithful.

I do.

winning the mustaprovince

DUSK HAS COME AND GONE by the time I get to the pumpkins, and I would ignore them and go inside, clean up, and eat—the time is after nine—except that I must seize an opportunity. A bloom will open in the morning and I cannot let the bees get to her before I do.

I take a flashlight and masking tape to the garden and search the wildly sprawling vines for the flower. The vine has many inflorescences in all stages of ripeness, and I am looking for a particular one. It is a female, set to open in about ten hours. Between the flower and the stem is an immature fruit, a miniature pumpkin-to-be. Up and down the vine, beneath the large, rough, white-spotted leaves, male flowers prepare to open in the morning. Then I spot the female.

I kneel down beside her, angling the light. Mosquitoes zero in on me quickly and circle, snarling and whining; one after another they dive-bomb. They are legion because of the rains, and relentless, and I wonder what the wild animals do after dark to endure them. I slap at mosquitoes as I tear masking tape and fold it over the blossom, shutting it tightly. I mark the blossom with a ripped length of blue cloth. To tie a bag over the blossom would be easier, but I have no pollination bags. I move among the vines and leaves and find a male blossom and repeat the procedure. Then another.

This pumpkin has a cool story. I was introduced to it at a small festival in the tiny village of Wardsboro, Vermont. The Gilfeather Turnip Festival celebrates the Gilfeather turnip—developed, most likely through hybridization, by John Gilfeather on his hillside farm in Wardsboro in the early 1900s. The festival is sponsored by the Wardsboro Friends of the Library, who sell packets of Gilfeather seed, locally designed T-shirts, and handmade Gilfeather cookbooks. Craftspeople vend their wares while local musicians wander around strumming. During the tasting hour the year I was there, I sampled caramelized turnips, turnip cake, turnip bread pudding, turnip soup, turnips with cheddar cheese.

A large glass jar at the registration table was filled with chocolate kisses. Whoever came closest to guessing how many were in it, a sign said, would win a pumpkin.

"Which pumpkin?" I asked a library volunteer at the table.

"That one." She pointed out one in a pile of pumpkins.

The pumpkin was as large and beautiful as a wheel of cheese. It was smooth, deeply ridged, and the color of apricots. It would easily make a dozen pies. I decided I was going to win that pumpkin. "May I count the kisses that I can see?" I asked the library volunteer.

"If you can see the candy through the glass, it's fair game," she said. "Counting is allowed."

Without touching the jar, I counted the kisses lying across the top. I counted approximate layers of kisses from top to bottom. I did some figuring. The jar was not perfectly cylindrical. The wider layers would have at least 12 extra kisses, and nine layers, more or less, were wider. I added 108 kisses for the wider layers. I figured some more.

What I've noticed with speculation is that guesses are usually too low. I cringe to think what this says about humans, that we are chronic underestimators. Knowing this, I added a hundred kisses to my total, wrote my guess—901—on a piece of paper, stuck my vote in the cardboard box, and turned on my hope machine.

"Did you see the pumpkin I'm going to win?" I asked my husband, who was selling his pottery upstairs in the library. When I talk like that, he believes me. He still thinks I'm magic.

"No, let's go see it."

I showed him the jar of kisses and the pumpkin. He reached for a slip of paper. "What was your guess?" he said.

"I'm not telling! And there's no need for you to bother guessing. I've won already." He scribbled on his paper, folded it, and slipped it through the slot of the box.

"I want that pumpkin," I said. "I'm going to save the seeds. Have you ever seen anything like it?"

"Can't say that I have."

At the end of the day, when volunteers counted the kisses, there were 891 in the jar. "Sweet," I thought. "I'm very close." The workers fidgeted through the entries, while I watched more nervously than I was willing to admit even to myself, and determined that the winning guess was 875. They sent someone off looking for the winner.

My guess was not 875. The pumpkin would not be living at my house. Then I did the math.

"Excuse me," I said. "I think you'll find another guess in the pile that's closer. 901." I was shocking myself at how greedy I had become. But it was an unusual and enthralling pumpkin. I wonder if George Washington Carver was like this.

"No," a volunteer said. "875 was the closest." I always think my accent may be a liability in situations such as this, because Southerners are often stereotyped as slow because their speech is unhurried. (When I asked a man on a Sitka street for directions once, he asked, "What, did you just fall off the turnip truck?")

"My guess was 901," I said. I could see energy finally reach a light-bulb behind the retiree's eyes. The volunteers had not thought of guesses exceeding the correct figure. "The rule is the closest guess, right?" I asked. "Not the closest that is less than?"

"That's correct."

The library ladies reexamined the guesses and found the 901. They confabbed among themselves. By this time someone had located the fabled winner among the crafts on the second floor and had brought him downstairs.

"Uh," one of the women said, "it looks as if we've made a mistake. Hold on just a minute. Is this you?" she asked me. "Did you guess 901?"

"I did," I said. "That's how I knew it was in the box."

"Well, that's definitely closer," she said.

"Yes, ma'am."

She turned to the nonwinner, for whom I had no sympathy. "We are so sorry," she told the man. "We made a mistake."

The man, who looked newly retired, was gracious. "It's not a problem," he said. He had a friendly face. That's what signaled the angel in me to emerge. "You can have the prize," I said, "if you really want it."

"Young lady, what would I do with a pumpkin like that?" he said.

"Oh, thank you," I said. Why was I thanking him? I'd won the damned thing. "I'd love to have it," I said. I knew exactly what to do with it. I planned to set it on the butcher block in my kitchen and photograph it and wish that I had grown it. I planned to tell the turnip festival story a hundred times. I planned to wait until the last possible hour next spring to cook it, maybe to wait even until a rotten spot appeared on it. I planned to bake it into pies and tarts.

I was smitten with the whimsicality of the scene, me winning an incredible pumpkin by guessing 901 chocolate kisses in a jar.

"What variety is it?" I asked the aides.

"Oh, the squash farmer told us. We have the name written down somewhere here," one of them replied.

"Squash farmer?"

"All she grows are squashes."

"Here's the name," another volunteer said. "Mustaprovince."

The name made no sense to me and after I lugged the pumpkin home, I promptly forgot it. A few weeks later, I moved the pumpkin to our cool basement, where it lasted a year without spoilage before I baked it into fabulous pies. I saved the seeds, but because its grower was a squash farmer, I was doubtful that the seeds were pure, meaning true to type.

But how I wanted to grow such charismatic, long-lasting, and delectable pumpkins. I needed three things. I needed the name of the pumpkin. I needed to know if it was an open-source variety. If so, then I needed seeds.

It just so happened that the next fall I attended Common Ground Fair, a huge outdoor organic agriculture show in Maine. By chance I spotted a pumpkin of the same variety in the exhibit hall. It was labeled MUSQUE DE PROVENCE. That was it! When I got home, I ordered seeds.

Now the seeds have grown into an insouciance of vines, and I am determined to produce pure seeds. I finish taping shut the last male flower, to prevent a wayfaring insect from haplessly contaminating the Musque de Provence's pollen with some other kind.

That night, just before falling asleep, I remind myself to pollinate the pumpkin flower first thing the next morning.

I sleep and I dream that I am taking care of a little girl. I find a goose egg and I am showing it to her when she drops it and it bursts, spilling a curdled yellow liquid that doesn't smell rotten. Then a small mother bird falls out of the shell, wet and unready for the world, followed by a baby bird, very tiny, swaddled in bits of hay. The two birds flounder on the floor. I wail softly, *Oh no no*, attempting to gather up the birds so that, although born immature the both of them, I might save their lives. The mother bird tries desperately to escape me, and as I try to cup her against a wall, she becomes a luna moth. Somewhere during the dream my erratic breathing wakes my husband. He says I've been holding my breath for ten to fifteen seconds at a time.

Next morning, I gently strip off petals and rub male anthers full of pollen onto the stigma of the female flower. Then I retape the female and wait. In a few days I see that the pollination is successful.

The blossom withers and drops away, and the fruit begins to enlarge. Over the weeks and months to come I keep vigil, watching and turning the pumpkin. I prop it on a board to keep ants and beetles from chewing on it. When it matures I will scoop out its seeds and dry them. I will give them to friends. I will grow more.

I may even become a squash farmer myself.

— 24 —

basic seed saving

To SAVE YOUR OWN SEEDS and get plants that are photocopies of the parents, you must grow open-pollinated seeds.

If you believe in moon magic, plant between either the last quarter and the new moon in the signs of Gemini for multiplication; or in the earthy signs of Cancer, Scorpio, and Pisces—believed to be the most productive constellations for aboveground crops. In many cultures, seed harvested at full moon is thought to have the best germinating power.

Select plants in your garden that have done well and have adapted to your temperament, soil, climate, and desires. If you want early melons, select seed from the earliest. If you want tolerance to cold, pick the plant that lives through the coldest night. You may be interested in disease resistance, late maturation, drought tolerance, or productivity. Keep notes if necessary. Mark your plants with tie-on markers (pieces of torn cotton cloth). Then choose from them the fruit whose characteristics most appeal to you.

Gathering seeds at the right time is important. For fleshy fruits, the seed is ready when the fruit is completely ripe. Flowering heads are tricky in that you must get the seed after maturity but before wind and animals scatter them.

My best advice to you, if you want to elevate your seed-saving interest to a passion or a scholarship and do it correctly, is to get Suzanne Ashworth's incredible book *Seed to Seed*, about which I once overheard someone say, "It seems so little to have all the answers." Ashworth knows (almost) everything there is to know about seed saving (and I added the *almost* only in case some small tidbit of information has not yet been discovered). The Seed Savers Exchange also periodically prints a seed-saving guide, which is an invaluable resource. I have the one that appeared in the *Seed Savers Summer Edition 1988*, a Seed Savers Exchange publication that served as their journal.

Maintaining seed purity is a science. You need to know how many of each variety to plant, how far varieties should be planted from each other, whether a variety is an annual or a biennial, how long the seeds are viable, and many more facts. You won't get many of those details from me here.

My goal is simply to plant a seed. In you.

Annuals

SELF-POLLINATORS

Some vegetables produce seed in one season and by reason of their botanical structure generally do not cross with others of their kind. This reproduction, called self-pollination, is easiest for the seed saver, since the seeds remain reasonably pure genetically without added protection from bagging or separating plants a great distance. Lettuce, tomatoes, peas, beans, and eggplant contain both male and female parts on the same flower (called a perfect flower). Their ovules are fertilized by their own pollen.

PEAS AND BEANS

In peas and beans, fertilization occurs before the flower opens. The anthers are snug against the stigma, ensuring pollination when the anthers release. These vegetables may be planted freely in the garden, although hard-core purists recommend separating beans by 150 feet or by another crop that will be flowering at the same time.

To Harvest Seeds: Let bean or pea pods dry on the plant until brown, then pick and shell. If cold weather looms, you can pull the entire plant and hang it upside down in a dry shelter. Label and store.

LETTUCE

Lettuce flowers occur like fireworks, in a bunch of little sprays which open over three to four weeks. Each tiny flower generates one lettuce seed. In regards to purity, to not tempt fate you should separate varieties of lettuces that will flower at the same time by 20 feet.

To Harvest Seeds: Seed heads will ripen in stages parallel to the timeline of the flowers, the first about eleven to thirteen days after the first bloom. The rule of thumb with lettuces is to harvest when about half the flowers on each plant have gone to seed. Cut the stalks of the flowers and make a bouquet, which you cram head-first in a paper bag and hang upside down until it is fully dry. Then the seed can be shaken or rubbed from the chaff. Label and store.

TOMATOES

Things start to get a little complicated here with the love apple. Most modern varieties of tomatoes are self-pollinating. They are bred to have short styles, with anthers that fuse together until the pollen has fallen on the stigma. The pollen then slithers down the style and fertilizes the ovules. Other varieties have long styles that protrude beyond the anthers. These are mostly heirlooms and in general are more likely to cross-pollinate. These tomatoes should be isolated by at least 100 feet—all varieties by at least 10 feet, ideally—because solitary bees have been known to transfer pollen between blooms. A flowering crop grown between varieties is always a helpful barrier.

To Harvest Seeds: See chapter 19.

EGGPLANT

Eggplants are mostly self-pollinated. To ensure purity, varieties must be separated by 50 feet or by caging—covering the entire plant with tight-woven cloth or screen in order to prevent entry by insects.

To Harvest Seeds: Eggplant should get very ripe, about to fall off the stalk. Let the fruit stay on the bush beyond the stage where it's edible. (Isn't it true that this phrase is a little vague? I've seen Dumpster-divers eat plenty of things that seemed beyond edible.) The color will turn dull and the eggplant will look sickly. I harvest the seeds by blending chunks of the almost rotten eggplant in a blender with water. Pour this mess into a bowl and, like tomatoes, viable seeds sink to the bottom. Eggplant seeds may be fermented like tomatoes to increase germination rates and kill seed-borne diseases, although it's not necessary. Strain, dry, label, and store.

Cross-Pollinators

Peppers and Okra
Although they have perfect flowers, these beauties are easily cross-pollinated by insects and should be kept 500 feet away from other varieties (a mile for okra) or, optionally, beneath screened cages—one variety to a cage. Okra flowers may easily be bagged.

To Harvest Seeds: Peppers turn red when they're ripe. Scrape the seed from the pepper core and dry out of the sun. The seeds are dry when a folded seed breaks in your fingers. For okra, pick fully mature pods and let them dry until they split open like a banana peel. Knock out the seeds. Label and store.

More Difficult Annuals

Squash, Cucumbers, Pumpkins, Cantaloupe, and Watermelons
These crops have separate male and female flowers and are outbreeding, meaning hardwired to cross-pollinate. To save their seeds and keep the varieties intact, you must do one of the following:

1. Learn to hand-pollinate. See chapter 16.
2. Keep seed stock separated by at least 200 feet (or a quarter mile for certain purity).
3. Plant only one of each species.

To Harvest Seeds: Wait until the fruits are fully ripe to pick them. Cucumbers will be yellow. Cut open, pick out the seeds, spread them on plates to dry. Label and store.

Radish
Radishes are wild beings that freely cross-pollinate. In fact, they need to cross-pollinate because they cannot fertilize their

ovules with pollen from the same plant. So the more plants you have of a variety, the better your pollination. You will not be able to eat the radish from the plants you want to save seed from, because they need that root, of course, to produce seeds. So leave the radish alone.

To Harvest Seeds: Watch it send up seed stalks that begin to flower. These turn into (edible) little torpedoes that slowly dry. Harvest, label, and store.

SPINACH

So much can be said about spinach. Remember, this crop is pollinated by wind, and it has male plants and female plants. Furthermore, it's hard to determine the sex of the plants until they've sent up a seed stalk. The best advice is to grow only one variety of spinach from which you plant to save seeds, keep double as many female plants as male, and strip off the seeds from the stalk right in the garden. At least this is what the books say. I have to confess that I've never saved spinach seed, so don't take my word for this. I have a hard enough time just growing it in southern Georgia. Oh, and the books also say that a thin fabric works fine to cage the spinach plants. Label and store.

CORN

Pollen from tassels of corn is swept long distances by wind and cultivars should be separated by time (early and late corn may be planted side by side) or a distance of over a quarter mile. You may hand-pollinate by shaking the tassels over the brand-new silks and bagging the pollinated crops.

To Harvest Seeds: Corn ears should harden on the stalk. Then bring inside, hang until dry, and shell. Label and store.

Biennials

These plants, which produce seeds in the second year of growth, include carrots, turnips, beets, kale, onions, parsnips, and salsify. The first year they produce a crop, which must be ignored (read: not eaten) and the plant must be maintained for a second year of growth. In northern climates biennials are dug up, overwintered in root cellars, and replanted the following spring. Firm types, like kohlrabi, are the easiest to overwinter; leafy types like collards tend to rot. If winters are mild, as ours are here in the subtropics of southern Georgia, biennials usually survive in the garden. For seed savers, most of these crops are self-sterile, require insects to pollinate, and cross-pollinate easily. All members of the brassica family (cabbage, broccoli, kale, collards, cauliflower, Brussels sprouts) cross with each other. If you're devoted to saving their seed, and I hope you are, you have to choose one cultivar from the entire family or isolate them by distance or screens.

To Harvest Seeds: Heading flowers are trickier to gather in that you must get the seed after maturity but before wind and animals scatter them.

Drying and Storing

Seeds must be thoroughly dry before storing. They should break, not bend. Life is triggered by moisture, and any droplet of water left in the seeds shortens their life span by keeping subtle life forces ticking away. A good rule is when you think seeds are dry, leave them another day. Temperatures over 110°F will damage seeds, so in hot climates they cannot be dried in direct sunlight. In humid conditions, subject them to a gentle heat—such as that from a solar dehydrator, a lightbulb, or a pilot light—kept around 90 degrees. Seeds that are prone to attack by weevils and other insect infestations also must be frozen in order to kill the eggs that have already been laid in the seeds. Store seeds under cool, dry conditions, since heat and humidity trigger germination and are enemies of viability.

In general, seeds should be stored in airtight containers, such as envelopes in coffee cans with lids taped airtight. Silica gel packets are often used for moisture control. Seeds last longest in the freezer if they are completely dry. If not the freezer, keep them in the refrigerator, if possible.

I have mentioned only a small percentage of the vast kinds of edible botanicals in the world that we will want to keep growing, for the sake of survival and diversity and pleasure, when the biotechs fail or when civil society gets strong enough to crush the multinationals—whichever comes first. For other crops, I suggest again that you get the Ashworth book or check online.

Are you confused enough already? Don't be. Seed saving is not hard. All you need is love.

— 25 —

seeds will make you a thief

THE HOLLYHOCKS in the Cimetière de Montmartre, blooming pink and white beneath the Basilique du Sacré-Coeur, near the headstone of Émile Zola, begged for their seeds to be stolen. The seeds came prepackaged, a doughnut of black flakes, like onyx, arranged in a ring, within a nubbin of calyx. To have these flowers growing in my garden would be to remember Paris and the beauty of its dead memorialized in marble and travertine. (Thank heavens I got to see Paris before I quit flying.)

I could not help myself. The seeds were so tempting, thrust toward me from the base of the hollyhock stem, itself yoked with more flowers, floating skyward like pink saucers. Hollyhocks bloom over an extended period of time, and their first blooms have turned to dry seed heads even as later blooms remain in bud. No one was around, not simply out of sight but not on guard anywhere, and the little packet came off so easily in my hand.

Mere hours later, Paris forgotten, Raven and I drove slowly through the south of France in drizzling rain. Southern France's rural countryside and its beautiful old villages, where houses were constructed of stone centuries ago, mesmerized me. A longing rose in me, to live in a landscape such as the one I passed through. I longed for window boxes, for a market where older women sat selling homemade melon jam and quail eggs, for a revolutionary spirit. I remember the purple and yellow plums snatched from tree boughs overhanging garden walls near a vineyard at Nitry; they were impeccably sweet. I think when I found the plums—so many, going unpicked, beaded with cool rain—I fell heedlessly in love with France. Before, I had been charmed. Now I was giddily, uselessly, mercilessly in love.

At one vineyard, Raven and I strolled in raincoats through a demonstration garden, which might have been our own except the labels were in French and some of the plants, like artichoke, we had never seen growing.

An aisle parted the garden in two, and rows angled off either side. Each weedless, perfect row was assigned to one kind of plant: peonies, beets, onions, carrots, strawberries, cosmos. One large fragrant herb had gone to seed, its umbel eye-level and pregnant, and I could not resist gathering a few of its seeds. A fennel, it still grows in my herb garden.

Along fence lines the roses were incredible, seemingly disease free and delicate, redolent, flawless, tender, raindrop-speckled, lovely. There were roses of all colors. I had never seen roses more magnetic than in France, and finally I understood why roses have inspired poetry from Elizabethan to modern. If roses grew easily from seeds, I would have stolen from them too.

I dried my modest bundles, wrapped in napkins and maps and pages photocopied from travel guides. There were the plum pits to care for, tied in a bandanna. I kept them as warm and dry as if they were little French poodles. Night after night, each spent in a different town, from a hotel in Semur-en-Auxois to a tent in the Alps to a goat farm in Jura, I guarded the seeds I gathered, little packets in my luggage, opening them to dry, closing them to travel.

I steal seeds in the hope of surrounding myself with a bewildering and awesome universe of plant life. But is it really stealing? The plant gives freely. If the plant is under ownership, in someone else's garden, is its reproduction then also someone's property? Aren't seeds, as Vandana Shiva argues, part of the commons? "The uniqueness of life is that it reproduces and that's the problem for capitalists," I heard her say once during a talk at Keene State College. "As if seeds pop out of corporate heads."

I am, of course, taking Shiva out of context. She is talking about the biotech industry's zeal to patent life-forms, including seeds, which is stealing from God. I'm not stealing from God. Or from the plant. Perhaps I am stealing from the person who planted the plant. Or who owns the property on which the plant grows. But the plant wants its seeds spread, and if they land halfway around the globe, all the better, from the plant's point of view. I'm an emissary of God.

— 26 —

gifts

Twelve wise women were sent out by their elder matriarch with pockets of seeds to replenish the world. When they reached my house, walking six abreast, the curtain of night had closed, and without words they entered the bare ground of my fields. They held things momentarily between their closed palms, lips moving silently, eyelashes laid gently in double brush lines. And then they lay down their gifts.

My garden, peaceful and calming as gardens are, has become a hotbed of activism, and sometimes a triage unit. To want to rescue anything is in my nature, although my husband doesn't understand why we can't eat everything we grow. We are saving those tomatoes for what? he asks. After many struggles—with moving from one place to another, and after we settled down, with bugs and disease, with drought and flood—I come to realize that I can only play a small part in a tragedy being played out on a world stage. I can only save so few things. My life is short, and time is precious.

My garden doesn't look typical. Growing in the garden are Moon and Stars watermelon, Fife Creek Cowhorn okra, Running Conch cowpea, Black radish, Green Glaze collard. My garden contains plants in all stages of life, from germination to going-to-seed. The barn hangs with feed bags of seed heads, drying, driving the mice wild: Hollow Crown parsnip, Outhouse hollyhocks, Long Keeper beets. The kitchen is stinky with seeds fermenting in their juices and waiting to be dried. Seeds proliferate in the freezer, in my office, in the seed bank, in the garden shed—in jars, credit card envelopes, coffee cans, medicine bottles, recycled seed packets. Our house looks like a strange fertility clinic, bent on reproduction, I a fertility goddess, hot to protect that which industrialism has bypassed and thereby made rare.

And what a beautiful and storied table we sometimes set. One gumbo is made with Hill Country Red okra, another of Long County Longhorn. Sometimes the dill pickles are sliced from common old Marketmores, and sometimes they are from Lemon. The steamed pod beans are Dragon Tongue or Pencil Pod Yellow or Black Valentine. Five Color Silverbeet chard gets sautéed with onion. The lettuce is Rouge de Hiver and Freckles, and the salad is topped with julienned Purple Vienna kohlrabi and Chioggia beets with their awesome concentric rings. The Coconut Squash soup started as a Gold-striped Cushaw.

At the farmers market we display the seeds at the front of our canopy. It is early April, garden-planting time in the South, although the climate has changed and the temperatures are too hot for normal. Raven has made a seed rack about the size of an old wooden Coke case, partitioned into fifteen sections. In it I have stacked my wallet-sized manila envelopes of seeds tagged with address labels. On each one is a little sticker of a flower or a peace sign. Inside each is a surprise and I don't mean the seeds.

One stack is Malabar spinach. Malabar is a spinach alternative that grows in Southern summers, on vines. It contains almost as much mucilage as okra, but the sliminess cooks out, and the leaves taste like spinach, they cook like spinach. Each packet contains one teaspoon of seeds.

A stack is chia, another is Jack bean. Four o'clocks fill a space next to lion's tail, Grandpa Ott's morning glory next to garlic chives.

The market is busy and Red Earth Farm is sandwiched between the strawberry farm and Georgia Southern University nursing students taking blood pressure. Mine is 102/66.

"Excellent," the young woman says.

Why shouldn't my blood pressure be excellent? I am a farmer, I am a seed grower, I am embedded in a local economy, I am among friends. Across from me Arianne and Elliot, young farmers of Hope Grows Farm, hawk their eggs and pastured poultry. I am sitting behind a table burgeoning with our crop of kale and chard, first of the season. I have been filling packets from jars of seed in a basket. A wrenlike woman in front of me is holding packets of seeds.

"Will this grow in a square-foot garden?" she asks. She holds up a packet of Bright Lights cosmos.

I picture a square foot of ground. The cosmos soar four feet tall, expansive, fractious. "They're unruly," I say.

"It's okay if they grow big," she says.

"Maybe plant one seed in the middle of a square," I say. I reach into my basket and bring out a fruit jar half full of seed.

"These are the cosmos seed," I say. I open the jar and shake a few of the pointy, sharp-ended shuttles into the bowl of my palm. "This tiny black crux is the seed. The rest is casing. I don't have the machinery to separate the chaff. Now you know the part to plant."

The second packet she has selected is lamb's-quarter, a wild edible that many people cultivate, of which a single plant will overwhelm a square foot of ground. It grows even more ebulliently than cosmos and produces leaves that taste like spinach. Lamb's-quarter, unlike spinach, loves heat, which we have plenty of. The woman thanks me and hands me two dollars.

An amiable man in his eighties is interested in cowpeas. He wants Purple Hull Pinkeye.

"I have some," I say. "But they're riddled with weevil holes and I can't sell them. I'll give you some and you can sort out the good ones." I fill a manila packet, slip in a surprise, and write the name on the packet. I also give the gentleman two kinds of cowpeas I packaged a year earlier that didn't sell. I've kept them in the refrigerator so likely they're viable.

A woman in jodhpurs and riding boots halts to ask if heirloom is the same as heritage.

"Yes, it is." I talk to her about heirlooms. She leaves with a packet of Hollow Crown parsnip and another of cilantro.

A young couple reports that they are tending their first garden. "We planted it yesterday," the man, short with buzzed, dark hair, says.

"We'd like to buy seeds," says the young woman. Her hair is dyed red as a neon sign. "But we planted all our space."

Instead, they buy one of Raven's pies and a jar of organic strawberry jam.

"Keep the garden watered," I say. "Especially until your seeds germinate. Dripping is better than sprinkling."

All morning, as the market progresses, my conversation consists of sentences like "Grow it on a fence" or "Plant it a quarter-inch deep" or "Sure, it would work for spanakopita." Of course the sinners sail by the seeds, but plenty of people pause. More and more people are stopping. They want to learn all they can about growing.

When the morning is over and our money is counted, this is what I know: Seeds are returning to circulation.

— 27 —

seed banking

IN 2008, Norway finished construction of a strange structure that reporters began to call the Doomsday Vault. Norwegians bored a tunnel into a solid-stone mountain in the permafrost on an island some seven hundred miles south of the North Pole and lined it with a meter's width of reinforced concrete. They, essentially, built a structure to last forever. They built it to withstand just about anything.

Why would Norway and its global partners build such a thing? To answer this question, we have to imagine scenarios that might precipitate the need to replenish foodstuffs globally. Suppose genetic engineering goes wild. Suppose a comet hits the earth. Suppose climate change rearranges agriculture as we currently practice it. Suppose seas rise?

The global seed bank was built to withstand even climate change. The tunnel was positioned high on a mountainside, 430 feet above sea level—130 feet higher than seawater is expected to rise in global warming's worst-case scenario, even if the polar icecaps melt. Tsunami waters won't reach it. Inside this remote and invincible mountain, the Norwegians are stashing seeds from all over the world, four million kinds of them.

There, in the Svalbard Global Seed Vault, the temperature is permanently below freezing, about 23°F. Refrigeration units lower the temperature further, to about -4°F. At this temperature, seeds stored in watertight and airtight foil packages last anywhere from fifty to two thousand years, depending on the type.

No matter what tomorrow brings—be it natural disaster or civil unrest, war or industrial accident or atomic bomb, or even genetic tinkering gone awry—the Norwegians hope their ark is impenetrable. "Seeds are not just seeds," said Jens Stotenberg, Norway's prime minister, speaking of the new vault and at least paying lip service to the idea of seed saving, "but the fundamental building blocks of human civilization."

The Doomsday Vault is a gene bank, which, as Cary Fowler (Seed Savers Exchange board member at the time and executive director of the Global Crop Diversity Trust, an organization that actively promoted the vault) has pointed out, is a fancy word for a freezer. Faced with a dwindling diversity of crop plants and their wild relatives, gene banks are coming into favor as bomb shelters for agriculture.

But the idea of a frozen vault in Scandanavia made one young farmer chuckle when he read about it. Daniel Botkin is cofounder of a three-acre family operation called Laughing Dog Farm in Gill, Massachusetts. In a blog entry he called the global gene bank "sexy." "The bigger task," he eloquently wrote, "is to decentralize society's entire relationship to agriculture, seeds, food production, and food security."

Gene bankers want seeds to continue as badly as do the home seed guardians. Seeds, however, cannot be kept forever like stacks of gold ingots. Seed must be conserved in situ, on farms and in gardens. They die otherwise. A foolproof seed vault in the Arctic Circle, where seed copies are kept, may get us through a catastrophe, but is not reliable for long-term genetic preservation. Scientists estimate that half of the seeds in the 1,400 seed banks worldwide are in desperate need of being grown out.

A garden is a living gene bank.

Gene bankers tend to be suspicious of gardens. When seeds are grown out they risk exposure to environmental pressures, such as a colder or warmer or wetter or drier growing season, which may force adaptations on a plant. Those being grown out are subject to the risks of cross-pollination. And, of course, a disastrous enough pressure, like a hurricane, and the seeds could be destroyed or lost altogether.

Gene banks are mostly interested in preserving genetic material. As Gary Nabhan explained in a 2005 interview with Arty Mangan of Bioneers, "Gene banks are genetic conservation projects." Gene banks do nothing to spread resources that, as Nabhan declares, were historically shared in systems of reciprocity. Traditional societies traded seeds to keep a variety going. For the Cherokee nation, for example, November was the Month of the Trading Moon (*nu da de qua*), a time of swapping between towns and tribes, and one must assume that the Cherokee, understanding the importance of trading in order to enlarge and enliven genetic resources, would swap seeds during the Trading Moon.

Gardeners, especially seed savers, are preserving names, stories, heritage, place, cuisine. Their aim is to retain the "culture" in "agriculture," rather than stripping it away, scientifically reducing it to mere germplasm. Gardeners want to regenerate seeds as often as possible, because seeds mean food and because gardeners often welcome adaptations.

The gene bank school of security is akin to people who think we'll be safer with a bigger military, more locks on the door, and a gun under the pillow. The Nabhan school of thought, on the other hand, believes that we're safer when we're out in the world, interacting with and attempting to understand the world and each other.

On a national level, the USDA maintains gene banks as part of a program called the National Plant Germplasm System (NPGS), whose mission is to acquire, preserve, evaluate, document, and distribute crop germplasm. However, the system is a network—as it freely admits—of federal, state, and *private* entities; "private industry underwrites selected projects," admits its website. If one reads the mission statement really closely, small farmers will not see themselves in it. NPGS is developing "new knowledge and technology," meaning that they believe that time-honored, simple, grassroots technology is not enough. NPGS believes in "a competitive food economy."

Mainly the NPGS doesn't bother with penny-ante growers like me; they deal with seed companies or university scientists who need stock for breeding. But recently I was able to talk the system out of another seed I'd lost.

Remember the Jack bean—that inch-long, eyeball-like bean, like a black-eye on steroids—that my grandmother had given me when I was a little girl? I could not find it in any source I checked. John Swenson told me that the bean was curated at the Southern Plant Genetic Resources Conservation Unit in Griffin, Georgia. "I am looking to obtain an accession of *Canavalia ensiformis*, commonly known as Jack bean," I wrote. "I see that you have this variety in your Taxa list. How could I go about obtaining this?" Brad Morris wrote back that he would need to know what type of research I was planning and the name of my organization.

I had a ruse ready:

I'd like to do some research on the possibility of using the bean as a mulch, either a living mulch, or perhaps the breakdown crop in no-till agriculture. Perhaps the bean could be scythed after growth is complete, so that it fixes nitrogen and also becomes the mulch that stays on the ground.

I want to experiment with it in a larger organic gardening system, such as a quarter-acre, to try to determine its value in the South in cover-cropping for both nutrients and water retention. Since it is native to India, I am interested in its potential in the southern United States. We are seeing dramatic changes in climate as a result of global warming and climate destabilization, and we are looking at possibilities for the future.

I am interested only in the white-seeded form of Canavalia ensiformis, *also called horse bean, Overlook, and sometimes sword bean. I will experiment with eating the young pods of the plant, although in large amounts the mature beans are said to be toxic. I also want to see how farm animals react to them as feed.*

In the end, I didn't write any of that. Did I really intend to experiment with our goats? Or my family? Instead, I wrote a simple request. "I am a nature writer and I am doing a project on heirloom seeds. Jack bean was the first bean my grandmother gave me when I was a kid. I would like to grow them again and I was given your contact info as a place to obtain them. Do I need to outline a study or is it enough that I am a writer and would like to grow these beans again?"

When the beans arrived, a little card with them said the accession had come from Costa Rica.

I have grown Jack beans every year since. Their story is relevant to my own. It helps define me. I'm exultant that a gene bank kept them alive during my own climate crisis, when the seas rose around me, and I'm pleased that it returned them to me when I was ready to take care of them again.

— 28 —

grassroots resistance

WE EARTHLINGS FIND OURSELVES at a crossroads, with our lofty ambitions grappling with environmental limits. The limits are so imminent that author and contrary farmer Gene Logsdon predicts that the most popular vacation spot of the future will be the backyard. The time is perfect, then, to become part of a beautiful uprising to maintain all options for feeding ourselves.

Every day I get news of resistance, not just the thousands upon thousands of gardeners who are quiet revolutionaries but the activists. For one thing, grassroots seed banks are springing up across the country and world. Most operate with a membership that grows out a plant, saves its seeds, and with them replenishes the bank's supply.

Charlotte Hagood and her friend Dove Stackhouse started the Sand Mountain Seed Bank in Alabama. Their bank is a collection of seeds that either originated in or naturalized to the Sand Mountain area of northeast Alabama and the tip of northwest Georgia. Bonnie's Best tomato hails from Union Springs, for example. The Sand Mountain Seed Bank, like most others, is not only a repository of the seeds of aging gardeners; it's a source where gardeners can get a start of legendary bioregional, open-pollinated varieties. Membership is inexpensive, ten dollars a year, because this is a homegrown operation. It's a labor of love.

Hagood laughed with me recently about her home state. "Alabama is so far behind," she said to me, "it's gonna be ahead when things crash. It has sort of a culture still here." She has been busily gathering up the remains of that food culture and keeping it alive until it is needed. The hope of Hagood and Stackhouse, who lead seed-saving workshops, is that once again White Half-runner beans and Choctaw Sweet Potato squash will be commonplace on the dinner table and that Alabamians will be proud of their food heritage. The seed bank provides a place "where the legacy of our ancestors can literally be kept alive," Hagood said.

On a larger scale, the Carolina Farm Stewardship Association has started One Seed at a Time, an organic seed bank devoted to Southeastern biodiversity. In addition, Southern Seed Legacy, currently housed at the University of North Texas in Denton, is banking seeds for the region. Southern Seed Legacy is the brainchild of two University of Georgia professors, Virginia Nazarea, anthropologist and author of *Heirloom Seeds and Their Keepers*, and her late husband Robert Rhoades. It too is a member-driven organization; a third of a member's seed harvest is returned to the bank and a third is passed along to another person.

For years Southern Seed Legacy operated out of the University of Georgia and was run mostly by graduate students. Every year they advertised a seed swap. One autumn Saturday, Raven and I traveled to one in steady rain four hours north, almost to Athens. We kept hoping that rain wouldn't be falling in Crawford, Georgia, where the swap was to be held at the Nazarea-Rhoades farm, Agrarian Connections. There, before his death, Rhoades collected historic farm buildings that were in different stages of restoration.

But it was raining, and in the deluge the swap was a bust. A few people erected colorful but soggy canopies and had small handfuls of seeds out to trade or sell. I walked around once, huddled under an umbrella. I walked around again. I perused Southern Seed Legacy's collection and watched an old couple drive up in a rickety truck and drop off sprouted garlic. I bought six tomatillo plants and we got back in our cars and drove the dismal four hours home.

In many ways, however, the seed swap was a gigantic success. At least a hundred people, many of them young, had come to check out the scene. Despite the nasty weather, some hardy believers were grilling barbecue for lunch. A band called the Roughbark Candyroaster Band was supposed to play. Everybody there was excited about genetic preservation. They braved even winter rains to be there.

Across the country, in state after state, bioregion after bioregion, the same thing is happening. People are standing up to guard seeds.

Seed banks have been given a novel twist in some areas of the country. These are seed *libraries*. Local seed groups deposit packages of seeds in racks or card catalog file drawers at public libraries. Library users are allowed to "check out" the seeds, the same way they would check out books, videos, or magazines. The check-out time is a growing season, and at the end of the season the patron "returns" seed to the library.

In some cases there are branch libraries at farms and at community centers. Volunteers stock the seed shelves and raise money to buy new varieties of seeds. They hold seed-saving orientations for new library patrons, not to mention seed-starting workshops and farm workdays. Some of them are part of the Transition Town movement and some are people who understand that systems we have relied on are collapsing. Some of them are stockpiling, others are attempting to identify varieties that do particularly well in their locales.

There's the Bay Area Seed Interchange Library (BASIL) in San Francisco and a very organized and impressive one in Richmond, the Richmond Grows Seed Lending Library, to name two.

When scientists at New Mexico State University announced plans to genetically engineer chili peppers in an attempt to make the industry more profitable, students organized a group called Occupy Green/Red Chile. The students are smart and well-spoken. They are determined. They're organizing petition drives and marches. "Everyone cares about this because in New Mexico chile isn't just a food, it's your culture," student Jessica Farrell told a reporter in 2011. "To secure the long-term protection of the farmers and the protection of consumers in terms of culture, there is no room for a genetically engineered seed."

Since March of 2011, residents of five small towns in Maine— Sedgwick, Blue Hill, Penobscot, Trenton, and Hope—voted to declare "food sovereignty" in their villages by passing "Local Food and Community Self-Governance Ordinances." State regulations favor industrial agriculture and have forbidden the sale of certain foods, like fresh milk or locally slaughtered meat. One ordinance proposed that "Sedgwick citizens possess the right to produce, process, sell, purchase, and consume local foods of their choosing."

"Tears of joy welled in my eyes as my town voted to adopt this ordinance," resident Mia Strong told a reporter. "I am so proud of my community. They made a stand for local food and our fundamental rights as citizens to choose that food."

The resistance takes many forms and the resistance grows. Way back in 1996, Greenpeace protestors sprayed milk-based paint on soybean fields near Atlantic, Iowa, where Monsanto research was taking place.

Outside the United States, a number of places, including Europe, require labeling. When GM foods are labeled, we'll know if people get sick after consuming them. States, including California and Vermont, are working to pass labeling laws. In the absence of labeling, activists design labels to attach illicitly to GM foods.

In 2006 Prince Charles set up the Bhumi Vardaan Foundation, a charity that works to end farmer suicides in India. Mother Seeds in Resistance of Chiapas, Mexico, is protecting indigenous corn from contamination by GM seeds. The Clif Bar Family Foundation awarded $375,000 in grants to three doctoral fellows pursuing organic plant breeding.

Guerrilla gardeners transform empty city lots overnight into gardens. Others make seedbombs, tight balls of wildflower seeds, and launch them onto highway medians, shoulders of sidewalks, and vacant lots in order to beautify their surroundings. Millions of people take seriously Patricia Klindienst's notion that "we eat our history and our politics every day," and they nourish themselves with organic, local, sustainable food.

The list goes on and on. This and more is what happens when we take the stewardship of food crops into our own hands.

Extinction is not an event, but a process. Extinction does not occur when the last germ of a certain seed loses its vitality. No, extinction occurs when a species can no longer evolve, a point called a genetic bottleneck. The loss of genetic resources—genetic erosion—both pauperizes and threatens human civilization. We are losing the plants that we have traditionally depended on, that built human society as we know it. Our food supply is in crisis, and to guard against catastrophe, either quick or prolonged, we need a good insurance policy. We need a bank account. We need a library card. We need quick action.

— 29 —

public breeding, private profit

MOST MODERN PLANT BREEDING takes place at government-funded experiment stations like the one where Randy Gardner worked for over thirty years. The guy's name makes me chuckle. I once knew an ichthyologist named Bass, and a chef named Baker. I've wondered about this, whether influence is at work here, or simply chance. When I visited, he had just retired as a tomato breeder at the Mountain Horticultural Crops Research Station, out on the interstate ten miles south of Asheville, North Carolina, a sprawling conglomeration of greenhouses, office buildings, and fields. Gardner doesn't seem to mind meeting me at the experiment station during late afternoon on a Sunday. He's still reporting to work as if he hadn't retired.

"I never had much aspiration that I would ever become anything in life," he said. "I planned on going into farming, since I was raised on a subsistence farm in Virginia." But he found himself at Cornell studying pomology, or fruit cultivation, and after graduating in 1976, Gardner took a job in North Carolina, four hours from home. Western North Carolina farmers were suffering from the wane of burley tobacco as a cash crop in the mid-1950s and were looking for another. The region's farmers were trying summer vine-ripened tomatoes, as opposed to tomatoes harvested when green. Breeding efforts for varieties adapted to green harvest centered in Florida and California. But North Carolina needed varieties bred specifically for its climate—days with temperatures in the mid-80s, nights in the 50s to 60s.

Six years after his arrival in Asheville, in 1982, Gardner released his first F1 tomato hybrid, a variety he named (as carefully as he named his children) Mountain Pride. "I get in mind what I want in a hybrid and then I develop parent lines and then I cross these two together," he said. "The first wilt resistance was developed in the 1950s. I went back to varieties from Florida and California that had wilt resistance and started there."

The work itself was done in large, sterile greenhouses and also in immaculate fields in the valley of the French Broad River. Mountain Pride, Gardner told me, was released openly to any seed companies that wished to produce seed. Castle Seed Company picked it up. Mountain Pride is still produced in limited quantity.

After Mountain Pride, funding for practical breeding programs got slashed and public research programs were forced to come up with their own budgets. "There's little grant money available to us researchers," Gardner said. "So there's no more open releases of hybrids. We give a hybrid exclusively to one company, with a royalty payback of 10 percent."

The exact parentage of the hybrid is usually a trade secret. Developing a new plant variety, then, is a bit like writing a book. The author's take is royalties. So what does this say for the future of food? I'd say it speaks to the need for public institutions to develop varieties with growers and eaters, not corporations, in mind.

What I couldn't understand was this: Why would a variety bred at a publicly funded research station by government breeders be sold to a private company? What is paid for by the public at public universities is then being patented and sold for profit. What does that do to the state of democratic and open scientific inquiry?

We know what it does.

Over his career Gardner developed twenty principal varieties. "That's a lot of breeding lines," he said. The two most popular cultigens are Mountain Fresh, for the main season, and Mountain Spring, for the early season. In 2003, Gardner released the grape tomato Smarty, which became instantly popular, to the Harris Moran Seed Company. He developed Sun Leaper, which sets fruit even at high temperatures, named after the late Paul W. Leeper, a Texas plant breeder. Gardner also released a series of plum tomatoes, including Plum Dandy, Plum Regal, and Plum Crimson.

Although retired, Gardner is still breeding, which is why he can't go fishing instead of showing up at work. This new project is a tomato with the positive attributes of an heirloom, like superior taste and nutrient content, but one resistant to early and late blight. This will allow, he said, the tomatoes "to have better shelf life so they can be marketed more widely than the local tailgate markets or for home garden production." He escorts me through a greenhouse to show me the current work. The greenhouse is large and contains potted tomato plants, all maybe a month old. Each of the plants is labeled. It's a different language, one I don't understand: X 056x 66 and 0 81 12 x 195. I'm curious how he pollinates tomatoes, I say, since they usually pollinate themselves before the flower even opens.

"We catch the flower before it matures and sheds pollen," says Gardner. "We use a highly technological tool to emasculate the flower." He grins quirkily and holds up a pair of tweezers. To emasculate means remove the anthers. "Then we pull out another very expensive, high-tech tool." He grins again and lifts an electric toothbrush, which happens to be nearby. "This is what we use to shake another plant's pollen onto the stigma."

Gardner and I squander a bunch of time looking around and it's almost sunset when we get out to the fields. Gardner begins moving down rows of tomatoes pendant with fruit. He explains that heirlooms are tastier because they have more foliage to feed the sugars in the fruit, which is why he's using heirloom germplasm in the development of these new hybrids. More sugars mean more carbohydrates.

"You're looking for high sugar/high acid," he said, "and lots of volatile materials for aroma. You want a tomato that has taste and smell at the same time." Smell enhances flavor through flavonoids, which are phytochemical compounds that serve as defense mechanisms in plants and that are associated with increased health benefits in humans.

I follow Gardner up and down the rows. He picks a tomato, cuts off a slice, offers me a taste, then tosses the rest to the ground.

"When I breed the way I do, it is not to select individual characteristics in plants," he says. "I look at the plant as a whole, the entire plant. That gets into the art of breeding. Of course, a lot of the art is based on science."

To get another perspective on breeding, I stopped to visit the trial gardens of Johnny's Selected Seeds, named after Johnny Appleseed.

Rob Johnston Jr., founder, is a breeder who has won six All-American Selections awards. He won in 1993 for Baby Bear Pie pumpkin, in 1998 for Bright Lights Swiss chard (a sensation in my garden), and in 2002 for the Diva cucumber (very rewarding to grow). Almost every year since then he has won national All-American Selections recognition for his work, most recently with squashes—Bonbon Buttercup and Sunshine Kabocha. He specializes in crops for fresh-market growers and big home gardeners, and thus many of his offerings are hybrids.

The day I visited the company's gardens Johnston was in the field, working with tomatoes. I had not made an appointment, but had simply found myself in the area and dropped in unannounced. Not wanting to bother Johnston, I spoke with his assistant, a young woman on a bicycle. It was a cloudy, cool, August Sunday in northern Maine. Rain looked imminent. At the trial gardens were a couple of buildings, then acres and acres of row crops, healthy and vigorous, greening the rolling hills. Greenhouses lined up one after the other.

Here Johnston and his workers grow commercial varieties side by side to compare them. Here they select specimens for seed saving. They segregate populations. They go to great pains to cross-pollinate varieties. They sometime hand-pollinate. They grow out seeds. They get dejected about a failure and excited about a promising creation.

When creating a new open-source variety, in addition to the seven years required to produce it, Johnny's Selected Seeds uses the eighth year to stabilize it. In the ninth year they grow stock seed, and in the tenth they farm it out to their seed growers. Then the seed can be sold.

"Do you ever put a lot of work into something and find that it's useless?" I ask the young woman.

"Not really," she says. "Usually something good comes out of it." I ask if I can just walk through the gardens. She says certainly, to make myself at home.

That humans need new varieties of crops is indisputable. As the environment changes, as conditions change, our crops must respond to those changes. The question I am mulling is, How should we be breeding new varieties—at public institutions or in private ones? Normally I would advocate that work on behalf of civilization be done in public facilities, supported by government money, with scientists trained to act on behalf of a country's citizenry. But when our institutions are controlled by forces outside government and when their products are snapped up by corporate interests and then sold back to us, supporting them becomes more difficult.

I would not normally advocate the development of new products by industry, since it concerns itself only with profit. In this case, however, I think some business interests are doing a better job of breeding new varieties than our public institutions. Johnny's Selected Seeds, for one, is producing some interesting and useful varieties. Johnny's is capitalism writ small. Johnny's employees own 66 percent of the company, with 100 percent ownership expected by 2015.

Who should do our breeding? I do not know the full answer, but I know part of it: the people who care first about life on earth. Will these be public or private? I do not know. For a moment I think we should consider the possibility that people will continue to rise up as they have for centuries, either self-educated or professionally trained, driven by a passion for plants. They will wear a mantle of service. Some of them, like Johnston or Tom Stearns, will start their own small companies. Inch by inch, row by row, they will work toward a true agriculture. They will improve our food increment by increment.

When I teach writing workshops, I often talk about our country's infatuation with stardom—how a few individuals, deserving or not, are hoisted into the national imagination. I'm talking about the Marilyn Monroes and the Leonardo DiCaprios. The cult of stardom comes at the expense of the entire culture, because civilization is advanced by many stars. Lots of them. When a single story obfuscates the many, the entire culture suffers.

The same is true in plant work, I believe. We need many gardeners, many seed savers, many breeders, in gardens large and small, working to make our relationship with plants more and more beneficial to all.

— 30 —

breed your own

THE FIRST I THOUGHT about being a plant breeder myself was at the headquarters of the Maine Organic Farmers and Gardeners Association (MOFGA) in 2008. I sometimes get invited here and there to talk about writing and the environment, and I happened to be in Unity, Maine, the day an evening talk about on-farm vegetable breeding was scheduled.

MOFGA runs an elaborate spread. On-site is a large timber-frame office and meeting space, demonstration gardens, and a complete homestead where some lucky apprentice gets to live for a year. Every September MOFGA hosts the Common Ground Fair, an organic conference, trade show-and-tell, and harvest fest. And it really is an agricultural fair. People still bring vegetables to enter in competition. I have listened to two judges talk about a certain squash not being true to its phenotype (what it's supposed to look like) and recommending the gardener check her seed supply. These are people who know their stuff.

About thirty thousand people show up to attend talks, visit booths, tour alternative energy projects, learn how to do things, and (not least of all) eat good food. No junk food can be sold by the purveyors. You won't find fried candy bars at Common Ground.

To not be cutting-edge would be difficult for a group of people as mobilized and organized as Maine's organic farmers. And the year I was there, they'd organized a workshop on vegetable breeding.

Jim Gerritsen of the Organic Seed Alliance, which works with farmers to grow high-quality seed, stands up. "We've got to be independent in our seed resources," he says. "We have to breed varieties designed to work under our local ecological conditions. What we want are locally adapted varieties suitable for organic growers."

There's that word *local* again. Locavore, local economics, and now locally adapted seed. This also means that if you're a seed saver, you're a seed selector, and thus a plant breeder, more or less, growing seed adapted to your locality. And as the seed adapts to soil, I heard a gardener at the Rodale Institute once say, the soil adapts to seed.

Gerritsen speaks for only a few minutes before introducing Bill Tracy, a plant breeder from the University of Wisconsin. "The astounding diversity of life on earth is directly related to natural selection," Tracy says. "This results in all the variations we see around us." A fundamental axiom of nature is that it creates diversity, and if growers select or breed seeds, they have contributed to diversity. Someone who's a seed saver is a minor deity. Tracy talks about the creative power of selection and applying selection pressure.

"This is so different from genetic engineering," he says. He said that selection is extremely powerful, usually remarkable and predictable, and is precise; seven cycles of selection can easily change a population significantly. To prove his point, he took a corn and produced from it a field corn, then while selecting from the same corn, produced a super-sweet corn. "The biochemists say that this is impossible," he says. "Fortunately I'm not a biochemist. All this variation we see around us is the result of selection giving us unexpected types."

Selection pressure is the difference between forcing a child to read or creating an atmosphere where a child wants to read. It's the difference between using maple syrup to sweeten pumpkin pie or using pumpkin-pie mix. It's two people talking at the edge of a field as opposed to them sending text messages via satellite Internet. Application of "selection pressure" is a grower easing a variety toward an imagined outcome, instead of forcing the outcome. It's human-scale—not industrial-scale—technology.

Tracy shows slides of work with a field corn he's developing with Martin Diffley, a Minnesota organic farmer (Gardens of Eagan). Diffley wants three traits—early vigor, weed competitiveness, and flavor. "The biggest thing you have to worry about with plant breeding is selecting for multiple traits," Tracy says.

I am truly astounded at the number of folk plant breeders in this country. It's like a national pastime. We earthlings are true plant lovers and no wonder. Diffley is one. Frank Morton of Oregon's Wild Garden Seed, who is also working with Tracy, is another. He calls seeds "the best deal in nature: dense nutritional matter with a self-organizing program and energy array. For cheap." He developed and introduced Wrinkled Crinkled Crumpled cress by crossing Persian and curled cresses and selecting from this gene pool. Morton also developed an open-pollinated, fast-maturing, vigorous sweet corn that germinates well in an Oregon spring's cold soil.

All this talk of the gardener participating in plant breeding led me to one of many agricultural epiphanies. Many people will disagree with me on this and in many ways I know they're right. What I realized, for myself, is that I don't have to grow a seed just because it has unique genetics. What I grow also has to produce and thrive. It seems hard-hearted, but one seed is lost forever and another is recovered or created. Sure, Will Bonsall was right, that I can't play God—but my personal job is to garden well, feed my family and community, and make a bit of a living at it. Someone else can save the genetically diverse but functionally weak cultivars. If an heirloom isn't working well, I believe, turn it into something that is.

Perhaps the most famous folk seed saver and plant breeder of our time is Glenn Drowns of Iowa. If he's not the most famous, he's no doubt the hardest working. I thought I was a hard worker until I learned Glenn's schedule. He works three full-time jobs. One of them is teaching middle- and high-school science, including chemistry and biology, in Calamus, Iowa. Another is running a mail-order business selling heirloom poultry and seeds, with a hundred-page catalog. The third is keeping those varieties and breeds alive at Sand Hill Preservation Center, outside Calamus.

I wanted to go visit Glenn, but I could not fathom using the fossil fuels to do so, and so I caught up with this busy man by phone on a Sunday, a rainy winter evening in Iowa that happened to be chilly but clear in Georgia. During our long and relaxed conversation Glenn explained his schedule. He's up at four thirty every morning to tend the poultry. A weather buff, he types in data about the day's conditions around seven o'clock, eats breakfast, and leaves for school at seven thirty. At four o'clock in the afternoon, when he's home again, he focuses on farm chores until dinner at six thirty. Evenings are spent filling seed orders, grading papers, and working on projects.

You have to hear the numbers to truly understand the dedication of Glenn to the diversity of food. He has single-handedly rescued poultry breeds from extinction, and now not only keeps alive 235 breeds but also raises and sells poults and chicks. Over the years, he has had as many as 2,000 plant varieties in his care, and even now he keeps many hundreds going—including 185 sweet potatoes, 200 corns, 150–200 squash, 700 tomatoes, and so on. The numbers are mind-boggling.

During the growing season Glenn is apt to spend two to three hours a day hand-pollinating squash, a crop about which he's passionate. "It's only work when it's not enjoyable," he said. "I've never been a person who can sit still."

A love of plants seems to have been born in Glenn. His mother found him planting seeds in her flowerpots when he was two. Plants were coming up everywhere, she later told him. By the time he was five she had to forbid him from bringing in citrus trees that he had planted in pots and cans all over their Salmon, Idaho, front porch. If Mrs. Drowns turned her back on Glenn in a store, she would find him at the seed rack, studying it.

"From the time I was a tiny child, I was gardening," Glenn told me—first with his next-door neighbor and then in his own plot. By nine years old, he had a plant business and was showing vegetables at the county fair. During high school Glenn worked on his first breeding project, a watermelon suited for short growing seasons, a development that is still on the market: Blacktail Mountain watermelon (available from Glenn himself, 70 days.) During his senior year Glenn came across an ad for the Seed Savers Exchange and sent away for information. "It was a whole new world," he said. "Now I was able to spend time with university specialists, which I hadn't been able to do before, to get my questions answered." Glenn went to Lewis-Clark State College in Lewiston, Idaho, where he continued to pursue his love of plants.

The summer he graduated from college, with a degree in biology and a certification to teach, Kent Whealy invited him to Decorah, Iowa, home of the Seed Savers Exchange. "You'll love it here," Whealy said. And Glenn did. "I fell in love with the fact that it was mile after mile of soil." Soon he had taken a teaching job and in 1988 bought land three hours south of Decorah, forty acres of a highly erodable sandhill made affordable by the fact that everybody who had tried to farm it had gone bankrupt. "It was solid sandburs, horse nettle, and blow sand," he said.

A quarter-century later, with liberal treatments of manure, compost, and green manure, "it's like night and day different."

Glenn has cut back on the number of seeds he maintains. "When I moved to Iowa in 1984," he said, "stuff was disappearing so rapidly." Now many wonderful seeds are being grown and saved by gardeners and seed companies, and the safety net for these heirlooms is strong. "There are a few other things I still want to do in life," he said.

"Like?"

"More plant breeding. I've always wanted to do plant breeding," he said.

"Why?"

The answer to that question harkens back to Glenn's childhood and his strange madness to garden. "I grew up as a child wanting to grow a butternut squash," he said. "In Idaho there wasn't a long enough growing season. There were varieties I didn't know were available and couldn't find." That was in the days before the Internet.

"How do you get an idea for a cultigen you want to develop?" I ask.

"I ask myself, how can I make two good things into something better?" he said. "Or more adaptable?"

"Give an example, please."

"I keep working on an open-pollinated sweet corn for cold climates. I've developed one that produces in forty-six days, which we've trialed as far north as Alaska. A gardener in Fairbanks got sweet corn."

"Have you named that one?"

"Yukon Supreme," he said. "I'm not that good with names."

"Sounds like a fine name to me. And besides the corn, what else?"

"I keep trying to improve the earliness and productivity of tomatoes," he said. "In addition, I'm working on a smooth-skinned version of the wrinkly Jimmy Nardello peppers, which are sweet and productive, but hold dirt when harvested." Jimmy Nardello Sweet Italian peppers appear on Slow Food USA's Ark of Taste. In 1887 they were brought to Connecticut from Basilicata, Italy, by Jimmy Nardello.

"I tend to get a few more projects than I can manage," Glenn said.

"Do you sell your developments?"

"No," he said. "And I never would. If I create something that makes the world a little easier, then I'm happy. My goal is to get as many unique things in the hands of people who'll benefit from them."

Glenn's greatest joy comes when he gets letters from people thrilled to have his seed. One woman ordered thirty packets of a pepper and

Glenn called her to ask if he could simply send pepper seed in bulk. "Oh, no," she said, "I want thirty packets because we're going to give them out at our family reunion. The pepper is an old family heirloom and I'm excited to see them still available." People are charmed to locate varieties they grew in their home countries. One man ordered five packets of a tomato that his great-grandfather, now in a nursing home, had developed in Texas.

An hour and a half passed quickly. Glenn worried that I was paying long distance charges and I told him that I had unlimited calling. I only had a couple more questions. I asked him about hope, were things getting better, were young people coming online who will keep food alive.

"I think that's the case," he said. "Otherwise, why would I keep beating myself up with a grueling schedule?"

In southern Iowa, Glenn is surrounded by the huge research farms of Monsanto. He is the only farmer in his county who is certified organic. The other day in the teachers' lounge at school Glenn brought out his lunch, which included broccoli. Monsanto had just announced that it had developed a broccoli with many times the nutrients of standard broccoli grown under similar conditions.

"Is that the new broccoli?" a teacher asked.

"No, this is definitely not super-broccoli," Glenn had said.

"I try to influence as many people as I can," he tells me now. "I tell people, agriculture doesn't have to be the way it is. There's a way to farm that's not destructive. Maybe that's why I'm here."

wheat anarchists

MAYBE IT'S THE ASHES that did it. Something turned Stephen Jones into a radical wheat breeder.

The ashes are those of William Jasper Spillman (1863–1931), the fifth wheat breeder at Washington State University, whose cremated remains were scattered in the fields where he labored. In his lifetime, Spillman warned against the industrialization of agriculture; as early as 1915 he wrote that tractors were too large. He coauthored *The Law of Diminishing Returns* in 1924, which said that if one input is increased while others remain static, overall returns will decrease over time. Fertilizer is one input. In terms of fertilizer, Spillman's law means that yield will not continue to climb with successive applications.

Stephen Jones, wheat breeder, has spent years of his life in wheat that grows from the ashes of William Jasper Spillman. Spillman is Jones's hero.

But the story is even more odd and complicated.

Spillman was born the exact month and year, maybe the same day, that yet another famous wheat breeder, Cyrus Guernsey Pringle, was being tortured for refusing to fight the Civil War. Pringle was drafted for service in July of 1863 and that fall was tortured for refusing to carry a weapon. Spillman was born in October.

Pringle was born in East Charlotte, Vermont, in 1838. As a young man he became intrigued not only with plants but with the nonviolent doctrine of the Quakers, and in 1863 was faced with a terrible test. Following his conscription into the Civil War, Pringle refused to perform all military duty. He was imprisoned and tortured, including being staked to the ground with his arms and legs outstretched in the form of an X. After a day of pain he reportedly wrote in his diary, "This has been the happiest day of my life, to be privileged to fight the battle for universal peace."

Pringle also wrote in his diary, upon entering Virginia, forced to carry a weapon he would not use: "Seeing, for the first time, a country made dreary by the war-blight, a country once adorned with groves and green pastures and meadows and fields of waving grain, and happy with a thousand homes, now laid with the ground, one realizes as he can in no other way something of the ruin that lies in the trail of a war."

President Lincoln personally petitioned Secretary of War Edwin M. Stanton to parole Pringle. After his release Pringle returned to his botanical work as a plant collector, nurseryman, botanical surveyor, and plant breeder. The first variety of wheat he developed he named Defiance.

When I first heard of Stephen Jones, I was listening to the panel of "On-farm Vegetable Breeding" experts at MOFGA. The minute this guy stood up I sat straighter in my seat. Jones is tall with bright blue eyes and a smile big enough to cause charley horses in his face. He stands up with a big, happy smile and tosses a little bomb in the room, which is full of growers who have just partaken of the most amazing potluck in the history of hippiedom, sitting comfortable and well-fed in our seats.

"Right now agriculture is centralized, globalized, and completely screwed up," Jones said. Not that Maine farmers didn't already know that. But this language isn't what they expected from a college man.

He switched on his slide show. "For ten thousand years on this planet we've had the right to save seeds for replanting," Jones was saying, "and now the biotech industries are working day and night to take that away. That's criminal and mind-boggling."

"Biotech is about ownership," he said. "That's all it's about. Owning the seed."

Jones is a dryland wheat breeder ("dryland" meaning not irrigated) formerly based in the Palouse, thousands of acres of what was once native prairie and what is now the rolling wheat fields of the Northwest, encompassing parts of Washington, Idaho, and Oregon. Jones comes from Washington State University (WSU), a land grant university that—like most land grants—typically and historically favors Big Ag.

Jones, however, who now directs the Northwestern Washington Research and Extension Center, located in the magnificent Skagit Valley just north of Seattle, believes in Small Wheat, small enough even for a sickle and hand-threshing. He also believes in local, in sustainable, in organic. (He attained organic certification for eleven acres on the WSU campus.) He's breeding wheat for low-input ag, organics, and nitrogen-use efficiency. Jones, who is on the board of The Land Institute, is also involved in converting wheat to a perennial crop.

He is not the kind of plant breeder who develops a variety and sells it to a company that then promotes it to farmers and returns a royalty to the breeder's institution. No, Jones takes seriously the idea of a public university and a public breeding program. He believes in the farmer as breeder, practicing what he calls evolutionary or participatory plant breeding. Defiance is his middle name.

The Washington Wheat Commission was pissed that Jones and colleagues went directly to wheat farmers and asked what traits they wanted in a wheat. They were so perturbed that in 2003 they staked Jones to the ground, metaphorically. The commission threatened to end its $1.66 million support for Jones's projects, mainly winter wheat development. He came under pressure because he refused to introduce herbicide-resistant wheat. The herbicide-resistant trait, called Clearfield, was owned by the firm BASF, which touts itself on its website as the world's leading chemical company.

"No, I don't enter into contracts with for-profit corporations," said the lone crusader. "I have a problem with public breeding programs not being public."

The story of wheat growing in this country has been the story of industrialization wildly triumphant. At MOFGA, I learned that in 1880, Maine, for example, grew forty thousand acres of wheat, producing 14 bushels to the acre. In the decades that followed, however, wheat-growing became chemical-intensive, concentrated, and machine-driven. The crop centralized in the Midwest and Maine fell off the USDA chart for wheat in 1946, when its production dipped below one thousand acres. Vermont went off the chart in 1931.

Washington produces the second-largest wheat harvest in the United States. In the 1920s and 1930s the Palouse of Washington produced high-yielding wheat, 100 bushels per acre—nonirrigated, nonchemical. The average chemical farm in Kansas today produces 36 bushels per acre. "We're gonna starve if we keep growing wheat in Kansas," said Jones.

According to him, growing wheat is easy (harvesting is not). Field trials these days are proving high yields in organic varieties. Madsen, for example, yielded 92 bushels an acre organic compared with 87 per acre chemical. Eltan grew out 115 organic and 105 chemical.

"Nobody can tell me we can't feed the world on organics," said Jones. His credo is:

1. Organic agriculture needs separate breeding programs. (In field trials he's found that the very worst wheat in organic can be the very best in chemical and vice versa.)
2. Farmers can breed their own damn varieties.
3. We need to diversify our fields and our science.

"Breeding takes time," says Jones. "But basically wheat breeds itself." Here are the steps.

1. Evaluate historical varieties. Not all made good bread or pasta.
2. Create variation. Select for it or let the environment select for it. Most important, utilize farmer knowledge and encourage farmer participation in the selection process. "I'm working with genetic wheat anarchists," says Jones. He means the farmers.
3. Harvest, replant, select.

One of Jones's success stories of evolutionary participatory plant breeding is young Lexi Roach, the granddaughter of Jim Moore, a organic wheat farmer in Kahlotus, Washington, whose farm receives only eight inches of rainfall a year. Lexi first became interested in wheat breeding when as a middle schooler she listened in on conversations that Jones had with her grandfather regarding breeding his own wheat. A few years later, she and her grandfather went to Western Washington University, where they crossed two varieties of dryland wheat that performed well on the Moore farm, producing new variation. For the next six years, they planted and stabilized the population. Lexi and her grandfather walked the rows together, pulling out weak plants and those with traits they didn't want. In 2007, yield from the Lexi 2 variety beat the farm's other top varieties by eight bushels per acre and in 2010 beat out fifty-nine other wheat varieties grown in a university-conducted trial.

In his last slide, Jones lists his research funders. This is the first time I have ever seen anybody do this—transparency and honesty. There is not a single corporation on the list. "No corporate influence," Jones says.

For a hero, Jones has Spillman. Spillman had Pringle. And I have Jones.

a vanishing plant wisdom

HOW MANY PEOPLE still know that huckleberries have seeds, and they have glands at the base of their leaves—and that blueberries do not? How many still know that the pungency of wild peppers is related to altitude and that the higher the altitude, the greater the pungency? How many know that the scapes of wild onion, like garlic, are edible? Or that a tobacco hornworm becomes a Carolina sphinx moth?

It strikes me that what we are doing at this point—in what I am hopeful is our evolution from an industrial society into a sustainable one, from the Cenozoic into the Ecozoic—is reclaiming lost decades in the garden. "Gardening is managing our relationship with nature," said writer and naturalist John Tallmadge. As biologist Robin Kimmerer said it, "We're all reading from the same book—the land. The library of knowledge is in the land."

Mostly we do not even know what we have lost. Most of us don't know that a pumpkin vine, for example, puts out flowers most commonly in a specific order of male and female. Many of us don't know what the male flower looks like as compared to the female. In fact, some of us don't even know that a pumpkin begins as a flower. Or where a pumpkin even comes from.

Part of the joy of this work is in discovery. Not long ago, when I was selling seed packets at the Statesboro Farmers Market, a couple in their second year of gardening, on a tiny scale, asked me about okra. "We grew some last year," they said. "We let it get big, then we couldn't eat it."

"Oh, no," I said. "Okra has to be picked young, when it's tender. The longer it grows, the tougher it gets, until it's inedible."

"We figured if we let it grow we'd have more of it."

"It's counterintuitive but true," I said. "You have to eat okra when it's young."

I recollected where I had learned the habits of okra. I had learned about it in any number of okra patches with which I had been associated as a kid, including that of Mr. Chavis, who lived in an empty building my dad owned and grew a patch of okra to sell. Someone at some point told me how to harvest okra, what size to cut. I also learned about okra in my mother's kitchen, watching her process the vegetable for stewing or frying and then helping her do it. If a paring knife did not easily enter the pod when pressed against it, the pod was too hard and must be thrown in the scraps pail for the hogs.

What other okra wisdom is there? It makes you itch when you pick it. You should wear long sleeves and gloves, unless you have grown Clemson Spineless (guess where that variety was developed). Okra is not harvested all at once, like a head of lettuce. Okra keeps producing. Keeping up with the harvest is difficult, since okra pours out pods. Every second or third day is the optimal harvesting time. You'll need a jackknife. Pick all pods that are ready. If you missed a pod during a previous harvest, cut it and throw it in the compost, unless you want to save it for seeds. If so, mark it with a string and leave it alone.

The job of a plant is to reproduce itself. When an okra plant throws some pods, and they are left to grow, and the seeds mature, that signals the plant to stop producing. Its work is done. But if you keep taking the fruit, the plant must keep making more, in hopes of completing its work on earth. An abundance of okra is a product of plant stress.

I don't know everything there is to know about okra. Nobody does. The important thing is that I want to learn as much as I can.

When Jane Howell sent me the Marriage garlic, the letter told me her plant wisdom. "This is a hard-necked garlic," she said. "It bears six to nine cloves around the hard neck. Then the small cloves will grow in size until they are big enough to make their own head." How many people still know that some garlic varieties are hard-necked and some are soft-necked? That rocket and wild rocket are not the same plant, although both belong to the brassica family?

It seems that every day I am an archeohomesteader, uncovering lost wisdom, and not just about plants: how to render fat, how to prevent scours in bottle-fed calves, how to caponize roosters. At the same time, I recognize that a burgeoning science enlarges the body of agrarian knowledge, some of it in response to modern challenges, and I eagerly add this information: where to find native earthworms to use in vermiculture so as not to spread invasive species, how to prune muscadines most effectively, how to control mites in beehives naturally. To plow forward without appreciating traditional wisdom, however, is a mistake. Virginia Nazarea writes about "connecting people to places through 'rivers of time' so that the present becomes full of possibilities and the future not so daunting." Without the traditional wisdom, learned in traditional ways, it's hard to move forward.

One area in which we are at risk of losing our wisdom is interspecies cooperation. Agribusiness, of course, favors monocropping, because that's what the machines can handle. But we are not machines, we gardeners.

Traditionally, we humans have been great at multicropping. In India, a second crop, such as fava beans, would be planted among wheat at exactly the right time—the wheat must not be so tall as to shade out the bean and the wheat must not be so short that the bean overwhelms it. When the wheat is cut, the fava beans scramble skyward.

In the Southern United States, vining cowpeas would be grown on cornstalks. The pea had to be the right variety, planted at exactly the right time. If planted too early, the pea overpowers the corn. If planted too late, it is shaded out. This explains why so many heirloom cowpeas carry the name "cornfield bean." My hundred-year-old neighbor, Leta Mac Stripling, told me that her family planted Velvet beans among their corn.

Companion planting, as well, encourages symbioses. Tomatoes and basil planted together benefit each other, the tomatoes attracting pollinators to the basil and the basil discouraging pests from the tomatoes. Strong-scented herbs like mint or rosemary help deter the cabbage butterfly. Bill McKibben reported from a trip he took to Cuba that, for some unknown reason, when green beans and cassava on the *organóponicos* are mixed in the same rows, yields improve 66 percent. Gardeners for centuries have been figuring out these things.

As Will Bonsall said to me, "Peasants were not stupid people."

I believe that we will relearn the ancient wisdom of the wild garden, and that we will become not only elders of the land but caretakers of it as well. If we get started learning from each other, learning from books, and (most imperative of all) learning from our own experiences on the land, we too will become wise.

Try marigolds near tomatoes. Horseradish with potatoes. Try pole beans growing on amaranth. Let sunflowers get up several inches then intersperse with pole beans. Try corn and White Tender Creaseback Cornfield beans.

stop walking around
doing nothing

"We have no place to start but where we are."
—WENDELL BERRY

I WANT ONE MORE TIME to remind you of the most powerful thing in the world. It is a seed. In this era of transition, between the Age of Industrialization and the Ecozoic Era, a seed is life. Because we don't know what is sealed in a seed, since the predetermined information is invisible, it can contain any number of surprises. Everything the seed has needed to know is encoded within it, and as the world changes, so it will discover everything it yet needs to know. That's the nature of adaptation and evolution, the two most important jobs we have on this planet. So even with the climate crisis, even with peak oil and soil, even with financial collapse, there will be seeds that possess all the information they and we need, which is why I think seeds are the ultimate metaphor.

Every morning I wake with fears and griefs; there are so many of them. I wake now into the news of storms. During the Cuban missile crisis, we built bunkers for fear of Soviet attack, where we would go to be safe. Now we build bunkers for seeds. When the storms have passed, what will we need to rebuild? We will need seeds. There is at least one in each of you. There is a bank of seeds within you. Let them grow.

I excused myself from a phone conversation the other day when I looked out the window and saw my husband trying to chase calves back into the pasture where they were supposed to be. They'd been separated during the day so that Raven could milk in the evening, and they wanted to get back to their moms.

"I gotta skedaddle," I told my friend. "The calves are out."

"I read this quote once," my friend replied, "that anyone who keeps animals will soon become slave to them." He elongated the word *slave*.

I'm not a quick thinker and I was outside, hollering like a banshee trying to cut a calf off and get her out of the young orchard, when I fabricated a witty retort: "Anyone who does *not* keep animals will soon become a slave to corporations."

The same with farming. Anyone who does not grow food will become a slave.

Agriculture has created in us a story-based, community-reliant, land-loving people. It has given us a head start on what I call the Age of Bells, the time when bells—cowbells, dinnerbells, bells of flowers— will again be ringing across the hills and plains. We are coming to the new age of agriculture better prepared: knowledgeable about growing, able to do with less, happy in our communities, firm in gender and racial equality, healthier. I believe that the organic and local-food movement is leading the way to re-creating cultures that are vibrant and vital. What we are witnessing in agriculture is no less than a revolution.

It also means we are on an edge—lots of edges, in fact. When I think of the edge, I think first of a literal one, the fencerow, which modern chemical agriculture has been destroying. This is the place where birds pooped out wild cherry seeds and wild cherry trees grew; and the place where, tired from the row, workers sat in the shade and told stories. It's where a lone farmer watched a mockingbird sing.

We occupy an edge between forest and field, the most exciting place in the world to me. We are on many edges: balancing the needs of the wild with the need to nourish people, balancing urban life with the need to eat, balancing concerns about human health with the need for productivity, weighing input against output, and making decisions based on both ecology and economy.

There is also a psychological edge we're all living on. We know that we're living in a world that is being devastated but also one replete with the beauty and power of life. We live on the boundary of deciding to make positive contributions although we know we are implicit in the destruction. We skate between apathy, because the truth of what's happening is painful to think about, versus action, any kind of action; and we skitter between the paralysis caused by grief and fear versus action. Every decision we have to make, whether it's a life-sustaining or a life-destroying one, is an edge. Our very psyches are on the edge, between dropping out and dropping in, between selling out and fighting back. Every single one of us.

The verge is a dangerous and frightening place. It's important to know that one is not alone on it. The edge holds a tremendous amount of ecological and cultural as well as intellectual power. I believe that we have to get comfortable with it.

How shall we live? As if we believe in the future. As if every one of us is a seed, which as you know is a sacred thing. In my wildest dreams the seeds of every species are speaking to me, calling out: *in all the bare spots on earth plant us and let us grow. On all the edges, plant seeds.*

One weekend in a storytelling session during a writing workshop, I asked participants to tell stories of hope. One man told about stopping on busy Highway 441 near Franklin, North Carolina, to rescue a box turtle that was only a foot from the yellow line. By the time the man turned around and got to the turtle, a woman in an SUV smashed it before his eyes. The next day the man saw another turtle, again on Highway 441, this time in the turn lane. The road was so busy that the man drove four miles before he had a chance to do a 180, and when he got back to the turtle, a white van had stopped in the suicide lane and had successfully rescued the reptile. "I'm not the only one," the man thought.

Another story was about ten-year-old girls at a summer camp who chose to feed a dead guinea pig to their large exotic snake; what was hopeful to the storyteller was the matter-of-fact reaction of the girls, a realness. Another story was about a professor's children who were visiting their dad on campus when they decided to make a sign: *Students, Stop Walking Around Doing Nothing.* They decorated the sign with pictures of the earth and peace signs.

What I really want to say is that these stories seem small compared to the enormity of the problems. Gandhi, however, pointed out that big problems need small solutions. Big problems need one courageous and willing person. Big problems need you doing what you desire to do and doing it with great authority, great knowledge, and great love. Maybe, just maybe, once you have picked up a tool at hand and started to work, someone will say of you, using a wonderful African proverb implying that someone is attempting something far beyond what is comfortable or maybe even possible for them: "She has gone in search of the fabulous birds of the sea."

Let's also get clear about hope. After talks I've been asked a hundred times—*Am I hopeful? How do I find hope? Do I stay hopeful? How?*

The assumption is that hope is a prerequisite for action. Without hope one becomes depressed and then unable to act. For many years I tried to say that I find hope in nature. Not long ago somebody asked me the question again and suddenly I thought, *Hope? Who needs hope?*

Do you feed your daughter because you have hope that she'll turn out okay? Hope is important to me, but I want to stress that I do not act because I have hope. I act whether I have hope or not. It is useless to rely on hope as motivation to do what's necessary and just and right. Why doesn't anybody ever talk about love as motivation to act?

Let me be even more truthful. It's not hope or love that keep me going. It's fight, which I will define as a life force surging in my heart.

So the question *How do I stay hopeful?* becomes as ludicrous as *How do I stay love-filled?* I'll tell you how. I wake every morning listening to the great-crested flycatcher call from the pear tree and I watch that fat old orange sun, always burning, rise flamboyantly over the pecan orchard. I watch Green Glaze collards go to seed. I watch hummingbirds in the red valentines of pigeon peas. Bottle-feeding the new calves after dark, I watch bats hunting insects. Before bed I walk outside and gaze up through the bare limbs of the swamp chestnut oak into the starry, starry sky above Red Earth Farm and I watch a meteor blaze a trail to earth.

I may not have a lot of hope but I have plenty of love, which gives me fight.

We are going to have to fall in love with place again and learn to stay put.

We are going to have to fall in love with each other.

We are going to have to learn courage and take action.

We are going to have to ignore that good ideas have been marginalized and rush them back to the center of attention.

One winter night I dreamed a marvelous dream. In it I watched a man jump from a plane. A colorful parachute opened against a tie-dyed, Technicolor sky. The air swirled with primary colors—in starbursts, vivid sundogs, spirals—like one of those six-inch-wide multicolored lollipops. Then I noticed a bicycle hanging from the parachute and I watched the man begin to pedal around through the universe. That was the totality of the dream but when I woke, I understood it to be a dream of possibility. We are leaping into the universe, and not only will we be given a parachute to save ourselves, we will be able to steer our course.

I am reminded of the quote from William Rivers Pitt, "The final truth is self-evident. You are the one you've been waiting for."

I say, It's the New Moon. Plant intentions. Don't burn them in a fire. Get really, really clear. It's going to be a powerful time. Sink into the place underground that seeds deserve.

I say, Rev up your awesome. Look around, so many people have put their shoulders into the load. You. Find a place to push. Pick up a tool—a hoe or a shovel. Start turning the compost bin, to make the soil in which the seed will grow. You will begin at the center, the center of many concentric circles that expand further and further out from you. You soon will become a local hero and a local rock star, and from there your influence will wash outward, even across the globe, where so many people are rising up like germinating embryos to claim food sovereignty, to rescue local seeds, and to guard human civilization's cornucopia. Come home. Have the courage to live the life you dream: There is nothing greater than this.

Many of our seeds have been lost forever. But we can protect what's left and in our revolutionary gardens we can develop the heirlooms of the future.

Begin now.

— 34 —

last stand

ARE YOU GOING to farmer up or just lie there and bleed?

disclosure

I DON'T OWN STOCK in any company mentioned in this book. This work is funded solely by Chelsea Green Publishing Company and the purchase of books by readers.

in memory

NOT LONG AFTER our visit to the farm of Jeff Bickart, his cancer returned in the form of brain tumors and treatment wasn't possible. Jeff had only time to get his things in order. He wrote a funeral service and assigned parts to friends. He collected his poetry and self-published it with the help of his friend Sylvia Davatz, because this manner of publishing is quick. Jeff's poems are brilliant and lovely. I heard that he sent a copy of the book to his mentor and hero, the poet Wendell Berry, and that Mr. Berry responded with a letter. Jeff used his last days to say goodbye to friends, to the farm, to the river, to his dreams, to a new variety of bean. With the finesse of someone expert at always finishing something, he completed his tenure on earth. He died October 17, 2008, at the age of forty-eight. As long as I am able, I will grow life-sustaining cowpeas in memory and honor of Jeff Bickart.

acknowledgments

BRIANNE GOODSPEED, my editor at Chelsea Green, breathed life into this book. Any writer should be so lucky as to have an editor such as she. I owe her untold gratitude for her incredibly careful read of the manuscript, her brilliant edits, and the friendship that developed along the way. Thank you, Brianne.

I thank all the staff at Chelsea Green, especially Margo Baldwin, Joni Praded, Melissa Jacobson, and Patricia Stone. Thanks to Ben Watson for his guidance. Intern Alaina Smith created the resource pages and we are grateful for her work. Kelly Blair is the person responsible for the lovely cover. I thank Eric Raetz for copyedits, Shay Totten and the entire marketing team for their creativity, and Jenna Stewart for setting up author events.

Thanks to Sam Stoloff, my agent at Frances Goldin Literary Agency.

My friend Larry Kopczak read a late version of the manuscript and provided an incisive and transformative critique. Plant pathologist Albert Culbreath directed me to sources I needed. In addition, I am grateful to the following people for reviewing sections of the book in order to make sure I stayed on the right path: Dave Brown, Dave Cavagnaro, Albert Culbreath, Jack Daniel, Sylvia Davatz, Glenn Drowns, Doug Elliott, Yanna Fishman, Randolph Gardner, Jim Gerritsen, Steven Jones, Woody Malot, Julia Shipley, Tom Stearns, Douglas Tarver, and Raven Waters. I am deeply grateful to all of you.

My one regret is that the writing of a volume such as this is never done. Always there are more seed savers, brimming with beautiful stories, that I want to interview and to write about. I could do another volume entirely. If you are not in this book, please believe me when I say that I wanted you to be, and please know that I would love to visit your garden and hold your seed collection in my hands. I thank all seed savers around the world for your work. I thank all who have given me seeds.

A number of people inspired and encouraged me with their own books and actions, and those people include Suzanne Ashworth, Wendell Berry, Gary Nabhan, Vandana Shiva, and Jeffrey Smith. Through thick and thin Susan Cerulean has been my literary confidante, personal sage, and dear friend for over two decades. Although not listed here, my many and cherished friends give me strength and surround me with love. I thank you all.

In addition, I am deeply grateful to Michael Cichon, M.D.; Susan Ganio, R.N.; Lee Arnold, P.A.; and all the people who devoted themselves to healing me from chronic Lyme disease. Their warmth and kindness sustained me through some rough days. I thank Elaine Cichon and the Clinic of Angels for financial support. I am very grateful to Stephen King, Margaret Morehouse, and all at the Haven Foundation; and Lisa Collier Cool and Trustees of the American Society of Journalists and Authors Charitable Trust for a grant from the Writers Emergency Assistance Fund. Without all of them, I would not be well and could not have written this.

Without my ancestors, all the way back to the dawn of time, I would not have a chance to witness life on this beloved, breathtaking planet and I thank them. I especially thank my grandmother, Beulah Miller Branch, from whom I received the gift of my first seeds, as well as my parents, Franklin and Lee Ada Branch Ray.

I thank my son, Silas, and my husband, Raven, for their love and their faith in me.

For both the quiet resistance of gardeners and the vocal resistance of activists, I thank you. May it grow.

what you can do

Eat real food.

Learn to cook it. If you are eating processed food, you are electing for agribusiness to feed you, and you will not be supporting the preservation of heritage seeds. Besides, cooking food is healthier for you. "Cooking outweighs class as a predictor of a healthy diet," said Michael Pollan.

Buy organic food. Organic regulations currently prohibit the use of GM.

Ask your local farmer or grocery store manager if the foods they are selling are GM.

Grow a garden.

Try to grow, between yourself and your friends, as much food as you consume. Make a trip to the supermarket a strange and intolerable experience.

Become a farmer.

Become a young farmer.

Become an elder farmer.

Become a girl farmer.

Become a small farmer.

Become an aspiring farmer.

Grow open-source seeds.

Buy seed from small, independent companies.

Buy organic seed.

Save your own seeds.

Trade seeds within your community.

Learn to hand-pollinate.

Select plants for seed saving based on your locality and conditions.

Learn to breed seed.

Never grow GM seed.

Nourish your pets and farm animals with non-GM feed.

Promote your local farmers market and farmers markets in general.

Become a seed activist.

Work for local and national sovereignty over seeds.

Work to make the United States a GM-free nation.

Work to refocus agricultural experiment stations.

Work to retrain extension agents in organic, seed-based, low-input systems.

When the Farm Bill is next up for reauthorization, work to have it represent small and organic farmers, not Big Ag.

Work for the rights of small farmers.

Work for the intellectual property rights of indigenous farmers.

Educate others about the importance of open-pollinated seeds.

Help pass a food sovereignty ordinance in your village, city, county, or state.

Succeed in passing laws requiring GM foods to be labeled in your state and country.

> Be joyful although you've considered all the facts.
> —Wendell Berry, "Manifesto:
> The Mad Farmer Liberation Front,"
> from *The Country of Marriage*

farmer rights

WE SUSTAINABLE FARMERS understand that we do not own life. We believe that farmers should have the right

to good food
to food sovereignty
to grow and share seed year to year, generation to generation
to the free exchange of genetic material among ourselves
to define our own agricultural policies
to choose diversity
to grow what we choose in the manner we choose without being
  subjected to chemical overspray, pollution residue, coal ash,
  and genetic drift
to sell what we grow, as long as it's safe, in the manner we see fit
to sell fresh, raw milk
to a food distribution system that does not displace families,
  farmers, animals, or wipe out indigenous peoples, landraces,
  and food customs
to offer leftover produce to gleaners
to be free of regulations sponsored by Big Ag designed to put us
  out of business
to be protected from sprawl and other development that
  threatens to swallow our farms
to economic security

broadcasters (resources)

ORGANIZATIONS

Biodynamic Farming and Gardening Association
> Biodynamics is a type of holistic organic farming.
> East Troy, Wisconsin
> www.biodynamics.com

Californians for GE-Free Agriculture
> A coalition of farmers and others committed to ecologically responsible and
> economically viable agriculture.
> http://calgefree.org/

Canadian Biotechnology Action Network
> A campaign for food sovereignty and environmental justice.
> Ottawa, Ontario, Canada
> www.cban.ca

Carrotmob
> Influencing businesses by spending instead of boycotting.
> www.carrotmob.org

Center for Cherokee Plants
> Saving the food crops of the Eastern Band of Cherokee Indians.
> Cherokee, North Carolina
> www.indiancountryextension.org/extension/program/center-cherokee-plants

Center for Food Safety
> Working to end harmful food production technologies and promote organic and
> other forms of sustainable agriculture.
> Washington, D.C.
> www.centerforfoodsafety.org

Cuatro Puertas
> The Arid Crop Seed Cache is one of the projects of this group that connects New
> Mexico's urban and rural economies.
> http://c4puertas.org

Educational Concerns for Hunger Organization (ECHO)
> Working with underutilized crops to improve the lives of the global poor, ECHO
> runs an amazing tropical fruit tree nursery and a seed bank.
> North Fort Myers, Florida
> www.echonet.org

Genetic Resources Action International (GRAIN)
> An international nonprofit working to support small farmers and social move-
> ments in the struggle for community-controlled and biodiverse food systems.
> Barcelona, Spain
> www.grain.org

Georgia Organics
>   Promoting organic agriculture in my state.
>   Atlanta, Georgia
>   www.georgiaorganics.org

Hawaii SEED
>   Protecting the Hawaiian Islands and their people from risks posed by genetically
>   engineered organisms.
>   Koloa, Hawaii
>   www.hawaiiseed.org

The Highland People's Food Seedbank Project
>   One of many such revolutionary projects around the world.
>   Inverness, Scotland
>   http://thehighlandpeoplesfoodseedbank.webs.com

Institute for Responsible Technology
>   Founded by Jeffrey Smith—first look at their ten reasons to avoid GM.
>   Fairfield, Iowa
>   www.responsibletechnology.org

International Seed Saving Institute
>   Highly recommended as the place to go first for seed-saving information.
>   Ketchum, Idaho
>   www.seedsave.org

Just Label It!
>   We have a right to know what's in our food.
>   Washington, D.C.
>   www.justlabelit.org

Label GMOs
>   The California ballot initiative.
>   www.labelgmos.org

Kerr Center for Sustainable Agriculture
>   Finding sustainable solutions for rural Oklahoma.
>   Poteau, Oklahoma
>   www.kerrcenter.com

The Keystone Center
>   Promoting Pennsylvania's regional food cultures and keeper of the Roughwood
>   Seed Collection.
>   Devon, Pennsylvania
>   www.williamwoysweaver.com

Kitchen Gardeners
>   An online community promoting food self-reliance.
>   Scarborough, Maine
>   www.kitchengardeners.org

Landless Workers Movement: Movimento dos Trabalhadores Rurais Sem Terra (MST)
>   Peacefully occupying unused land.
>   Brazil
>   www.mstbrazil.org

Laughing Dog Farm
  Daniel Botkin offers great information on his website.
  Gill, Massachusetts
  www.laughingdogfarm.com
La Via Campesina: International Peasant Movement
  Representing 150 organizations opposing corporate agriculture;
    a grassroots mass movement.
  Jakarta, Indonesia
  www.viacampesina.org
Maine Organic Farmers and Gardeners Association (MOFGA)
  If you live in or near Maine, join this force now.
  Unity, Maine
  www.mofga.org
Magsasaka at Siyentipiko para sa Pag-unlad ng Agrikultura (MASIPAG)
  Farmer-led network working to maintain local control of genetic and biological
    resources and agricultural production in the Phillippines.
  Los Baños, Laguna, Philippines
  www.masipag.org
Midwest Organic and Sustainable Education Service (MOSES)
  Organizes a huge organic farming conference.
  Spring Valley, Wisconsin
  www.mosesorganic.org
Nebraska Sustainable Agriculture Society
  Organic food is on the table in Nebraska.
  Ceresco, Nebraska
  www.nebsusag.org
Northeast Organic Farming Association (NOFA)
  Keeping the Northeast at the head of the pack.
  Stevenson, Connecticut
  www.nofa.org
Organic Farming Research Foundation
  Using scientific research to transform agriculture.
  Santa Cruz, California
  http://ofrf.org
Organic Seed Alliance
  Advancing the ethical development and stewardship of agricultural seed.
  Port Townsend, Washington
  www.seedalliance.org
Ozark Seed Bank
  Historically meaningful seeds for Ozark farmers.
  Brixey, Missouri
  www.onegarden.org
Populuxe Seed Bank
  You can start one too!
  Victoria, British Columbia, Canada
  www.theseedbank.net

Primal Seeds
  A response to industrial biopiracy and monocults.
  www.primalseeds.org
Rodale Institute
  Going strong in organic agriculture since 1947.
  Kutztown, Pennsylvania
  www.rodaleinstitute.org
Sacred Seeds
  Creating and supporting plant sanctuaries that address the rapid loss of
    biodiversity and cultural knowledge.
  http://sacredseedssanctuary.org
Sand Mountain Seed Bank
  Homegrown seed preservation.
  Albertville, Alabama
  www.sandmtnseedbank.org
Savannah Urban Garden Alliance
  Making this town near my home a food oasis, via community gardens.
  Savannah, Georgia
  www.sugacentral.org
The Seed Ambassadors Project
  The path to locally adapted seed in the Willamette Valley. Check out
    their seed-saving zine.
  Sweet Home, Oregon
  www.seedambassadors.org
Seed Matters
  Research and education.
  Emeryville, California
  http://seedmatters.org
Seed Savers Exchange
  Nationwide members exchange open-pollinated seeds of vegetables, herbs, and flowers.
  Decorah, Iowa
  www.seedsavers.org
Seedy Sunday
  Huge community seed swap in the United Kingdom.
  Brighton and Hove, England
  www.seedysunday.org
Slow Food USA
  As opposed to fast food.
  Brooklyn, New York
  www.slowfoodusa.org
South Carolina Crop Improvement Association Foundation Seed Program
  This germbank includes a heirloom list, many grown for display at South Carolina
    Botanical Gardens.
  Clemson, South Carolina
  http://virtual.clemson.edu/groups/seed/

Southern Foodways Alliance
   Documenting, studying, and celebrating the South's food cultures.
   Oxford, Missouri
   www.southernfoodways.org
Southern Seed Legacy
   A seed exchange that concentrates on the South.
   Denton, Texas
   http://pacs.unt.edu/southernseedlegacy/
Sustainable Mountain Agriculture Center Inc.
   Saving yesterday's seeds for today and tomorrow, as well as economic development
      for the rural Appalachians; great source for heirloom beans.
   Berea, Kentucky
   www.heirlooms.org
Whole Grain Connection
   Wheat seed project for California (a local-wheat movement).
   Mountain View, California
   http://sustainablegrains.org
Wild Farm Alliance
   Integrating farms and ranches into natural landscapes—one effort is toward
      farming practices that accommodate wildlife.
   Watsonville, California
   www.wildfarmalliance.org
World Wide Opportunities on Organic Farms (WWOOF)
   Volunteer on farms around the world in exchange for room, board, and an education.
   www.wwoof.org

PRINT AND ONLINE PUBLICATIONS

100 Heirloom Tomatoes for the American Garden
   Carolyn Male, Workman Publishing, 1999
Acres USA Magazine: The Voice of Eco-Agriculture
   Austin, Texas
   www.acresusa.com
"Becky and the Beanstock: Rebecca Pastor's Heirloom Bean Blog"
   St. Louis, Missouri
   http://beckyandthebeanstock.com
Botany for Gardeners (third edition)
   Brian Capon, Timber Press, 2010
*Breed Your Own Vegetable Varieties: The Gardener's and Farmer's Guide to Plant Breeding
   and Seed Saving*
   Carol Deppe, Chelsea Green, 2000
Breeding Organic Vegetables: A Step-by-Step Guide for Growers
   Rowen White and Bryan Connolly, Northeast Organic Farming Association of
      New York, 2011
Crops and Man
   Jack R. Harlan, American Society of Agronomy, Crop Science Society, 1992

Cultures of Habitat: On Nature, Culture, and Story
Gary Paul Nabhan, Counterpoint, 1997
Dave's Garden
Encyclopedic information submitted by a large, international community of gardeners.
El Segundo, California
http://davesgarden.com
The Earth Knows My Name: Food, Culture and Sustainability in the Gardens of Ethnic Americans
Patricia Klindienst, Beacon Press, 2006
The Ecologist (especially the November 2008 issue)
London, UK
www.theecologist.org
Gardening with Heirloom Seeds: Tried-and-True Flowers, Fruits, and Vegetables for a New Generation
Lynn Coulter, University of North Carolina Press, 2006
Gathering: Memoir of a Seed Saver
Diane Ott Whealy, Seed Savers Exchange, 2011
"GM-Free Churches: A Project of the Berea Gardens Agricultural Center"
www.gmfreechurches.blogspot.com
Green Food: An A-to-Z Guide (Sage Reference Series on Green Society: Toward a
Sustainable Future)
Dustin R. Mulvaney, Sage Publications, 2010
Growing Garden Seeds: A Manual for Gardeners and Small Farmers
Robert Johnston Jr., Johnny's Selected Seeds, 1976
A Guide to Seed Saving, Seed Stewardship, and Seed Sovereignty (fourth edition)
The Seed Ambassadors Project, 2010
www.seedsavers.net
*Heirloom Seeds and Their Keepers: Marginality and Memory in the Conservation of
Biological Diversity*
Virginia Nazarea, University of Arizona Press, 2005
*Heirloom Vegetable Gardening: A Master Gardener's Guide to Planting, Seed Saving, and
Cultural History*
William Woys Weaver, Holt, Henry & Company, 1997
Hybrid: The History and Science of Plant Breeding
Noel Kingsbury, The University of Chicago Press, 2009
Pleasant Valley
Louis Bromfield, Wooster Book Company, 1945
*Renewing America's Food Traditions: Saving and Savoring the Continent's Most
Endangered Foods*
Gary Paul Nabhan (editor), Chelsea Green, 2008
"Seed Matters: Cultivating a Sustainable Food Future"
www.seedmatters.org/blog/
Seed to Seed: Seed Saving and Growing Techniques for Vegetable Gardeners
Suzanne Ashworth, Seed Savers Exchange, 2002
Seedling, GRAIN's Quarterly Magazine
Barcelona, Spain
www.grain.org/article/categories/15-seedling
*Seeds of Change—the Living Treasure: The Passionate Story of the Growing Movement to
Restore Biodiversity and Revolutionize the Way We Think about Food*
Kenny Ausubel, HarperCollins, 1994

Seeds of Deception: Exposing Industry and Government Lies about the Safety of the Genetically Engineered Foods You're Eating
Jeffrey M. Smith, Yes! Books, 2003
www.seedsofdeception.com
"Small Is Beautiful (and Radical)"
Elliot Coleman, *Grist*, February 3, 2010
www.grist.org/article/small-is-beautiful-and-radical
Soiled and Seeded: Cultivating a Garden Culture
An online garden magazine intent on expanding the conventional approach to gardens and the practice of gardening.
www.soiledandseeded.com
Stolen Harvest: The Hijacking of the Global Food Supply
Vandana Shiva, South End Press, 2000
Uncertain Peril: Genetic Engineering and the Future of Seeds
Claire Hope Cummings, Beacon Press, 2008
Vegetable Seed Saving Handbook
http://howtosaveseeds.com
"The Wheel of Life: Food, Climate, Human Rights, and the Economy"
Debbie Barker, Center for Food Safety (published by the Heinrich Böll Stiftung), September 2011
www.centerforfoodsafety.org/wp-content/uploads/2011/09/TheWheelofLife_Barker.pdf
"Why Iraqi Farmers Might Prefer Death to Order 81"
Nancy Scola, *AlterNet*, September 2007
www.alternet.org/world/62273/
The World According to Monsanto
Marie-Monique Robin, New Press, 2008

OTHER

The Iraqi Seed Project
Film by Emma Piper-Burket exploring the history and current trials of agriculture in Iraq, seeking solutions to restoring Iraq's farmland after years of war.
www.iraqiseedproject.com
Our Seeds: Seeds Blong Yumi
Film by Michel and Jude Fanton celebrating traditional food plants and the people who grow them, shot in eleven countries.
www.cityfarmer.info/2012/03/30/our-seeds-one-hour-documentary-released-on-the-net/
Saving the Seed: The Struggle for Food Sovereignty in Honduras
Film by Scott Turner and Claire Kane Boychuk (while students at the University of North Carolina–Chapel Hill).
http://savingtheseed.org
Seed Hunter
Film by Sally Ingleton, Philip Bull, and Tony Stevens.
www.seedhunter.com/index.html
Sin, Salvation, and Saving Seeds: Live at Seed Savers' Heritage Farm
Music CD by Greg Brown

A Few Suppliers

A Few Good Plants
  El Dorado Springs, Missouri
  www.afewgoodplants.com
Abundant Life Seeds
  Cottage Grove, Oregon
  www.abundantlifeseeds.com
Amishland Heirloom Seeeds
  Reamstown, Pennsylvania
  www.amishlandseeds.com
Anson Mills
  Columbia, South Carolina
  www.ansonmills.com
The Ark Institute
  Easton, Pennsylvania
  www.arkinstitute.com
Baker Creek Heirloom Seeds
  Mansfield, Missouri
  www.rareseeds.com
Bountiful Gardens
  Willits, California
  www.bountifulgardens.org
Family Farmers Seed Cooperative
  Williams, Oregon
  http://organicseedcoop.com
Fedco Seeds
  Waterville, Maine
  www.fedcoseeds.com
Gourmet Garlic Gardens
  Bangs, Texas
  www.gourmetgarlicgardens.com
High Mowing Organic Seeds
  Wolcott, Vermont
  www.highmowingseeds.com
Hudson Valley Seed Library
  Accord, New York
  www.seedlibrary.org
J. L. Hudson, Seedsman
  La Honda, California
  www.jlhudsonseeds.net

Johnny's Selected Seeds
  Winslow, Maine
  www.johnnyseeds.com
Sand Hill Preservation Center
  Calamus, Iowa
  www.sandhillpreservation.com
Skyfire Garden Seeds
  Kanopolis, Kansas
  http://skyfiregardenseeds.com
Sierra Seed Co-op
  North San Juan, California
  www.sierraseeds.org
Southern Exposure Seed Company
  Mineral, Virginia
  www.southernexposure.com
Sustainable Seed Company
  Petaluma, California
  www.sustainableseedco.com
Territorial Seeds
  Cottage Grove, Oregon
  www.territorialseed.com
Terroir Seeds
  Chino Valley, Arizona
  www.underwoodgardens.com
Turtle Tree Seed
  Copake, New York
  www.turtletreeseed.org
Uprising Seeds
  Bellingham, Washington
  http://uprisingorganics.com
Wild Garden Seed
  Philomath, Oregon
  www.wildgardenseed.com
Wood Prairie Farm
  Bridgewater, Maine
  www.woodprairie.com

index

Harlan, Jack, 71
Harris Moran Seed Company, 171
Heald, Paul J., 5
health hazards, 7–8, 37–40
heirloom seeds, 23–24, 28, 35, 51–56,
    58–64, 71, 121–8
Heirloom Seeds and Their Keepers
    (Nazarea), 71–72
Heritage Farm, 101–9
Hi-Bred Corn Company, 11
High Mowing Organic Seeds, 61, 106
Hoedown Organic Farm, 27–32
hollyhocks, 156
Hope Grows Farm, 159
Hopi amaranth, 135
horizontal production, 35–36
Howard, Albert, 47–48
Howell, Jane, 73–74, 187
hummingbirds, 94–95
hybrids, 11–13, 60–62, 171–2

imperfect flowers, 86
indeterminate *vs.* determinate
    tomatoes, 123
industrial agriculture, 10–16, 46
inflorescence, 86
*International Treaty on Plant Genetic
    Resources for Food and Agriculture,* 108
Iraq, 33–35
Irwin (friend), 29, 47–48
Isenbarger, Jill, 62

Jack beans, 23–24, 164–5
Jefferson, Thomas, 103
Jimmy Nardello peppers, 179
Johnny's Selected Seeds, 29, 62, 172–4
Johnston, Rob, Jr., 173
Jones, Stephen, 181–5

kamut, 42–43
Kasper, Lynne Rossetto, 105–7
Keener, Bill, 138–42
Keener corn, 138–42
Khadighar Farm, 57–64
King, F.H., 46
King of the Early beans, 81
Klindienst, Patricia, 169
Kte'pi, Bill, 36

labeling, 38–39, 169
Lamarck, Jean Baptiste, 4
lamb's-quarter, 160
Laughing Dog Farm, 163
Leeper, Paul W., 171
lettuce, 151–2
locally adapted seed, 65–69, 175–6
Logsdon, Gene, 166
Longhorn/Long County okra, 77–78
Lysenko, Trofim, 4

Maine Organic Farmers and Gardeners
    Association (MOFGA), 175
maize, 5, 12
Malabar spinach, 159
Mangan, Arty, 163
marriage garlic, 73–74, 187
McCullars White Top Set onions, 104
McKibben, Bill, 188
Mitla Black tepary beans, 82
MOFGA. *See* Maine Organic Farmers
    and Gardeners Association
monocultures, 8, 10, 188–9
Monocultures of the Mind (Shiva), 72
monoecious plants, 86
Monsanto, 14, 15, 36–37, 115–20, 169
Moore, Jim, 184
Moore, Michael, 44
Morris, Brad, 164–5
Morton, Frank, 41, 177
Mother Seeds of Resistance, 169
Mountain Fresh tomato, 171
Mountain Horticultural Crops Research
    Station, 170–2
Mountain Pride tomato, 170–1
Mountain Spring tomato, 171
Mull Kidney beans, 82
Musque de Provence pumpkin, 148
mutagens, 13

Nabhan, Gary Paul, 41, 107, 163
National Organics Program, 39
National Plant Germplasm System
    (NPGS), 164
natural selection, 176
Nazarea, Virginia, 71–72, 167
New Mexico State University, 168
nonparticipation, 2

about the author

WRITER, naturalist, and activist Janisse Ray is a seed saver, seed exchanger, and seed banker, and has gardened for twenty-five years. She is the author of several books, including *Pinhook* and *Ecology of a Cracker Childhood*, a *New York Times* Notable Book. Ray is on the faculty of Chatham University's low-residency MFA program, and is a Woodrow Wilson Visiting Fellow. She has won a Southern Independent Booksellers Award for Poetry, a Southeastern Booksellers Award for Nonfiction, an American Book Award, the Southern Environmental Law Center Award for Outstanding Writing, and a Southern Book Critics Circle Award. She attempts to live a simple, sustainable life on a farm in southern Georgia with her husband, Raven Waters.

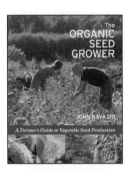

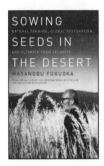

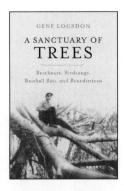
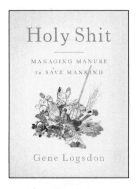

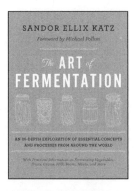